Sustainable Refurbishment

Sustainable Refurbishment

Sunil Shah
Director
Acclaro Advisory

A John Wiley & Sons, Ltd., Publication

Blackwell Publishing was acquired by John Wiley & Sons in February 2007.
Blackwell's publishing program has been merged with Wiley's global
Scientific, Technical and Medical business to form Wiley-Blackwell.

Registered office:
John Wiley & Sons, Ltd, The Atrium, Southern Gate, Chichester, West
Sussex, PO19 8SQ, UK

Editorial offices:
9600 Garsington Road, Oxford, OX4 2DQ, UK
The Atrium, Southern Gate, Chichester, West Sussex, PO19 8SQ, UK
2121 State Avenue, Ames, Iowa 50014-8300, USA

For details of our global editorial offices, for customer services and for information
about how to apply for permission to reuse the copyright material in this book please
see our website at www.wiley.com/wiley-blackwell.

Library of Congress Cataloging-in-Publication Data
Shah, Sunil, 1974–
    Sustainable refurbishment / Sunil Shah, Director, Acclaro Advisory.
      pages cm
    Includes bibliographical references and index.
    ISBN 978-1-4051-9508-9 (pbk. : alk. paper)   1. Sustainable buildings–
Design and construction.   2. Buildings–Repair and reconstruction.
3. Green technology.   I. Title.
    TH880.S53 2012
    690.028'6–dc23
                                                2012006682

A catalogue record for this book is available from the British Library.

Wiley also publishes its books in a variety of electronic formats. Some content that
appears in print may not be available in electronic books.

Set in 10/12 pt Optima by Toppan Best-set Premedia Limited

Cover image courtesy of iStockPhoto
Cover design by Meaden Creative

Contents

Preface

The question of how to refurbish a building in a sustainable manner has been asked frequently over the past few years, but increasingly has also come with a question of how this can be measured and achieved. And that is where the difficulties started – and where the originations of this book came from.

Refurbishments are becoming increasingly recognised as a valid alternative to new build after many years of being seen as simply a way to cheaply extend the life of a building. Historically, the focus has been placed upon new build as a way to develop additional space or the regeneration of areas that have fallen out of favour. New build benefits from starting with a clean slate and with a structured building process requiring planning approval, design and building standards to be met, and the introduction of operational strategies to suit the investor. Refurbishments are altogether more complicated with a structure and services already in situ and requiring adaptation. It is also arguably a more complex and challenging type of project to get right.

Legislation, corporate responsibility and the reduced access to capital for investment have all contributed to the increased level of refurbishments taking place, together with technological improvements to enable us to better utilise existing buildings. This latter point has helped bring down costs to refurbish, but also has the ability to provide Grade-A rentable space back to the market.

At the same time, the lessons learnt from new build about integrating sustainability measures in the design and, importantly, in the operational phase of the building have not been transferred across. There are a number of factors contributing to the lack of strong sustainability measures in the refurbishment market. The lack of a recognised definition for a refurbishment has certainly caused difficulties when understanding the return on investment for a variety of sustainability measures – as plantroom changes through whole-floor changes and 'taking back to shell' all constitute a refurbishment of a building but will have very different approaches.

A major gap is the relatively little experience in delivering sustainable-refurbishment projects, which are commonly delivered as a one-off project, rather than as a major programme moving forwards. The lack of experience brings with it a host of contributing factors, such as materials that are not cost-effective due to economies of scale, lack of exemplar projects, and a reduced level of understanding by the client and design community as to how to refurbish in a sustainable manner.

An obvious omission concerns the social aspects related to refurbishments. Occupant performance and behaviours are covered, but the wider community benefits that can be realised from such a project have not been captured. This has been a difficult decision and one that was made due to the relatively limited amount of robust information in this area. A significant amount of work is taking place in this area with a view to better understanding how the community can be proactively engaged and brought into any refurbishment project, thus enabling social inclusion.

This book has been separated into three sections to respond to each of these challenges. The first section provides a framework to integrate the lessons learnt from sustainable buildings into the refurbishment market together with exemplars. The middle section reviews the carbon agenda within buildings, including the material to review how a low-carbon design and operational philosophy can be used. The final section provides a review of the environmental considerations that can be incorporated, including material choice, biodiversity enhancement, water use and transport.

Sunil Shah
May 2012

Acknowledgements

There are a number of people and organisations who have helped in writing this book.

Firstly, I couldn't have written this book without the support of my family who have reviewed drafts and provided some sanity during the past months.

Specific thanks go to Ian Shaw, David Walker and Dave Cheshire for their guidance and use of information to help set the context of the book.

Also the patience and support provided by the team at Wiley-Blackwell who have provided gentle encouragement throughout the process.

Lastly and most importantly, I have to thank those clients and organisations who have provided the case studies and stories which provide the backbone to this book, without which the reality of what can be achieved could not have been put forward.

Abbreviations

ASHRAE	American Society of Heating, Refrigeration and Air Conditioning Engineers
B(E)MS	Building (Energy) Management System
BAP	Biodiversity Action Plan
BCO	British Council for Offices
BRE	Building Research Establishment
BREEAM	Building Research Establishment Environmental Assessment Methodology
BS	British Standard
BSRIA	Building Services Research and Information Association
CFC	Chlorofluorocarbon
CHP	Combined Heat and Power
CIBSE	Chartered Institution of Building Services Engineers
CIRIA	Construction Industry Research Association
CO_2	Carbon Dioxide
CR	Corporate Responsibility
DJSI	Dow Jones Sustainability Index
EBIT	Earnings before Interest and Tax
EC	European Community
ECJ	European Court of Justice
EMS	Environmental Management System
EPA	US Environmental Protection Agency
EPBD	Energy Performance of Buildings Directive
EPC	Energy Performance Certificate
EU	European Union
FSC	Forest Stewardship Certification
FTE	Full Time Equivalent
GABS	Global Alliance for Building Sustainably
GBTool	Green Building Tool

GDP	Gross Domestic Product
GFT	Global Fortune
GHG	Greenhouse Gas
GRI	Global Reporting Initiative
HCFC	Hydrochlorofluorocarbon
HK-BEAM	Hong Kong Building Environmental Assessment Method
HVAC	Heating, Ventilation and Air Conditioning
IAQ	Indoor Air Quality
ILO	International Labour Organisation
IPPC	Integrated Pollution Prevention and Control
IPCC	Intergovernmental Panel on Climate Change
ISO	International Organisation for Standardisation
IT	Information Technology
KPI	Key Performance Indicator
kW	Kilowatt
kWh	Kilowatt Hours
LCA	Life cycle analysis
LEED	Leadership in Energy and Environmental Design
LPG	Liquefied Petroleum Gas
M&E	Mechanical and Electrical
M&T	Monitoring & Targeting
MEPI	Measuring Environmental Performance of Industry
MRF	Materials Recycling Facility
NABERS	National Australian Building Environmental Rating System
NGO	Non Governmental Organisation
NIA	Net Internal Area
NPI	Normalised Performance Index
O&M	Operation and Maintenance
ODS	Ozone Depleting Substance
p.a.	per annum
PEFC	Programme for the Endorsement of Forest Certification
POE	Pre-/Post-Occupancy Evaluation
PR	Public Relations
PV	Photovoltaic
RICS	Royal Institution of Chartered Surveyors
ROI	Return on Investment
SA	Social Accountability
SAP	Sustainability Action Plan
SBS	Sick Building Syndrome
SCP	Sustainable Construction Potential
SME	Small and Medium Sized Enterprise
SRI	Socially Responsible Investment
SWMP	Site Waste Management Plan
TBL	Triple Bottom Line
UK	United Kingdom
UN	United Nations
UNEP	United Nations Environment Programme
UNFCCC	United Nations Framework Convention on Climate Change
UPS	Uninterrupted Power Supply

US	United States
USGBC	US Green Building Council
VFM	Value for Money
VOC	Volatile Organic Compounds
WC	Water Closet
WEEE	Waste Electrical and Electronic Equipment Directive
WFD	Waste Framework Directive
WLV	Whole Life Value
WRI	World Resources Institute
WSSD	World Summit on Sustainable Development
WWF	World Wildlife Fund

Part 1

Introduction to Building Refurbishment

This opening part will provide an introduction to the refurbishment of buildings, detailing the scope of what is meant by the subject. It will provide a definition of refurbishment covering different types, but will not cover the reasons why an organisation may choose to refurbish their building, assuming that this decision has already been taken.

The churn rate of buildings is less than 2 per cent in many countries, with economic downturns focusing attention on maximising the use of existing buildings. This, in turn, has increased the rate of refurbishing various buildings, but operating at varying levels from a 'refresh' through to a more fundamental upgrade.

Sustainability will be introduced as an umbrella term with a series of key issues under this umbrella that affect the refurbishment of buildings. This will be reviewed against the 14 themes of sustainability, which are used and referenced throughout this book.

This book has been written to enable a number of the technical aspects to be read in conjunction with the sector-specific examples and will provide the structure and linkage, together with an overview as to the direction that sustainable refurbishment is taking with regard to regulatory, reputational and financial aspects.

Developers often face the dilemma of whether to refurbish or redevelop. If sustainable construction is to be widely adopted as a practice, developers need to be convinced that it is commercially viable, including both capital costs and whole-life costs. Refurbishment is perceived to be the more sustainable option, but this is not always the case.

Buildings should have second and third uses. Many buildings since the 1960s are incapable of being changed, which limits their potential both as an asset and also to develop a cohesive community landscape.

Sustainable Refurbishment, First Edition. Sunil Shah.
© 2012 Sunil Shah. Published 2012 by Blackwell Publishing Ltd.

1 What is Building Refurbishment?

The traditional building cycle sees the built asset typically designed to last sixty years and undergoing a variety of changes prior to being demolished and a new facility commonly arising in its place. There are many buildings which have stood for much longer than sixty years and, in some cases, their external aspects are protected as listed buildings; however, they are often changed internally to reflect the modern-day demands of the working environment and the market. Likewise, there is an increasing trend for assets to be sweated over a relatively short lifetime of twenty or thirty years to maximise returns.

In all these scenarios, there is an element of building refurbishment that will take place. As can be seen from these examples, however, the scale, type and level of detail of the refurbishment are hugely variable.

Chapter Learning Guide

This chapter will provide an overview of building refurbishment, its growth globally and the challenge in trying to define its role to meet the changing business requirements.

- Definitions of building refurbishment;
- Size of market and global differences;
- Occupant satisfaction providing the link between productivity and well-being that can be influenced;
- Growth and changes affecting refurbishment;
- The future refurbishment and where the industry is going.

(Continued)

Sustainable Refurbishment, First Edition. Sunil Shah.
© 2012 Sunil Shah. Published 2012 by Blackwell Publishing Ltd.

> Key messages include:
>
> • There is an increase in the level and scale of refurbishment taking place across a range of buildings;
> • Building refurbishment is a standard approach to revitalising facilities to extend their life.

1.1 Introduction

The refurbishment of buildings is a long-standing practice dating back centuries as parts of stately homes, castles and churches have been extended and remodelled to reflect the wealth and architectural style of the owners at the time.

Whilst this aspect of the building lifecycle has been around for many centuries, it has not gained recognition in the recent past whilst failing to capture the imagination. With the recent growth in buildings and new construction, the refurbishment of the older stock has taken a back seat. Alongside this, there have been a loss of knowledge and inherent understanding of how to refurbish buildings in a sustainable manner across the various organisations involved in the built environment. This has resulted in many of the lessons needing to be relearnt.

There is an implicit recognition that many of the interventions we are now making through 'retro-fitting' are 'experimental'. This is because the performance of innovative technologies, or new combinations of tried technologies, can be predicted only with limited confidence. The need to monitor eventual performance, especially under occupation, is paramount if we are serious about providing effective refurbishment. Evaluating building performance needs to be holistic and stretch beyond the narrow concerns of the building as a machine, to embrace a more organic (realistic) view of the built environment as the platform on which people live their diverse lives. This sentiment is discussed further in Chapter 3.

This book does not look at the decision whether or not to refurbish an asset, but is based upon the premise that the refurbishment decision-making process has concluded. Such decisions are complex, but might include the following areas:

• What does the financial modelling support?
• Which option delivers the best increase in lettable area?
• What does the local market require?
• Does the developer's portfolio favour a particular option?
• Can the existing building be refurbished to a competitive level of quality?
• Is there an end-user preference for a particular option?
• Is there more than a marginal difference in new-build construction cost versus refurbishment cost?

Refurbishment is a large and ever-present element of construction workload, and one that becomes more important in a downturn. The challenge is to extract value out of tired commercial buildings through targeted investment. Refurbishment work is, by its very nature, diverse, ranging from redecoration to total reconstruction based upon a retained structure. What all refurbishment projects share in common is a greater risk profile than the equivalent new-build project and, therefore, risk allocation is critical.

1.2   Definitions of Refurbishment

It is surprisingly difficult to find definitions of building refurbishment in the publications of the various professional bodies looking after this area, which may explain the additional lack of information on the subject.

Webster's New World College Dictionary[1] provides the following definition:

re·fur·bish (ri fur'bish)

transitive verb

to brighten, freshen, or polish up again; renovate
Etymology: re- + furbish

Synonyms: <u>awakening</u>, face-lift, facelifting, <u>rebirth</u>, recharging, recommencement, refilling, reformation, regeneration, rejuvenation, renovation, reopening, replenishment, <u>restoration</u>, resumption, <u>resurrection</u>, revampment, revitalization, <u>revival</u>

The US Green Building Council provides the following definition in terms of major and minor refurbishments which will dictate the type of assessment methodology to be applied[2] (see Section 2.2 for more details):

'. . . a major renovation involves elements of HVAC renovation, significant envelope modifications and major internal rehabilitation.'

Likewise, the BRE provides the following for its assessment methodology approach[3]:

'. . . a major refurbishment project is a project that results in the provision, extension or alteration of thermal elements and/or building services and fittings.

- Thermal elements include walls, roofs and floors.
- Fittings include windows (incl. rooflights), entrance doors.
- Building services include lighting, heating and mechanical ventilation/ cooling'

In both cases with BRE and USGBC, changes to buildings that do not affect the thermal performance or significant envelope changes are considered to be minor and part of the operational management of the building.

However, this is a simplistic approach whereby changes to a building that can affect the asset value are not captured under the criteria of a refurbishment categorised by the BRE or LEED. Typical areas include cosmetic changes to common areas such as the reception or lift lobbies involving decorating. Such 'refresh' examples have an effect in making the building more attractive to tenants and improving the occupants' conditions. Whilst consisting of fairly small changes, the 'refresh' market does represent a significant part of the refurbishment market.

An alternative approach of reviewing refurbishments within a building looks at the scale of the changes being made: from a light touch through to a fundamental alteration taking the building back to its shell. This kind of approach will cover and include a whole range of upgrades, refurbishments and retro-fits made to a building regardless of the scale and size of the change being made. The main basis of this is that all of these areas can be classed as refurbishments and there will be scope to improve the level of sustainability within these changes.

The British Council for Offices (BCO) considers four options that represent degrees of intervention into existing buildings[4]. These reflect many concerns raised as to how much, or how little, should be done to create a product of value that achieves the developer's objectives. The BCO approach concerns the extent of the redevelopment, the value this will achieve in the long run, the necessary levels of both internal and external improvement and how to avoid spending more than necessary to maximise return on investment.

A similar approach has also been postulated by BSRIA and the BRE based upon five levels including a final level which promotes demolition. The first four levels, however, broadly correspond to the levels of intervention captured by the BCO.

The section below provides a series of levels to define refurbishment, which will be used throughout the book, to help identify when sustainable refurbishment criteria are more applicable to some levels.

Table 1.1 captures the level of refurbishment necessary for a given building based upon a matrix combining the building's performance and its condition. Those buildings that perform well in both areas require relatively limited change to maintain market value or increase the level of their sustainable performance. Those buildings that score poorly in both areas will require a significant change. However, the levels are blurred, rather than absolute, as the number of factors influencing the choice of refurbishment needs to be included in a decision.

Level of Refurbishment

- Level 1: Light Touch/Refresh
- Level 2: Medium Intervention
- Level 3: Extensive Intervention
- Level 4: Comprehensive Refurbishment
- Level 5: Demolition

Full details of each of the five levels of refurbishment are provided below:

Table 1.1: Level of refurbishment

		Building Condition			
		Excellent	Good	Poor	Very Poor
Building Performance	Excellent	Maintain	Level 1	Level 2	Level 3
	Good	Level 1	Level 2	Level 3	Level 3
	Poor	Level 2	Level 3	Level 3	Level 4
	Very Poor	Level 3	Level 3	Level 4	Level 5

Level 1 – Light Touch/Refresh

This represents the lowest investment, and delivers the least opportunity to generate value from the potential improvements to the building, but is a quick and relatively unobtrusive approach. The scope of works includes decorating, changing carpet tiles, replacing ceilings, repairing and upgrading minor elements of the building, including servicing the building's plant. Office floorspaces are often the most heavily used areas and are particularly suited to the light-touch approach as less effort is required to improve them. Externally, little will change and the immediate impression will be that the building has undergone routine maintenance rather than a refurbishment; however, reception and entrance rebranding can easily be included within light-touch improvements and can have a significant impact on staff, visitors and potential tenants.

Typically, the light touch will be applied when tenants leave a building, or to refresh a tired owned building for customers and visitors and therefore its focus is on the areas most visible.

Level 2 – Medium Intervention

This would include the scope of works outlined in Level 1 plus the replacement of building services in part of the building, cores, reception upgrades and a revised workspace strategy. A 'medium intervention' would see the public areas and office floorspaces given a more significant overhaul with the replacement of materials, fixtures and fittings. This could include replacement of toilet sanitary ware, new lighting, reception floor materials and entrance features. Replacement facilities for teapoints and upgrades to communications room facilities can also be achieved. The level of refurbishment is likely to be limited to works that fall below the threshold where a Building Regulations, or an equivalent such as ASHRAE, application would be necessary.

Such an approach is more likely from an owner-occupied building where the facility forms part of the image of the organisation and retaining staff and therefore the refurbishment is a means to enhance this. In conjunction, owner occupiers have a greater potential to refurbish at any time rather than wait for leases to end or plan for disruption.

Level 3 – Extensive Intervention

This would include the works outlined in Level 2 plus a full replacement of building services, some building-fabric changes, possible extensions to the floor plates and the remodelling of cores and communal areas. The enhancements should be carefully considered to commit only to the most appropriate improvements necessary to meet current Building Regulations, or an equivalent such as ASHRAE, standards and to 'future-proof' the building for a further 15 – 20 years.

An 'Extensive Intervention' delivers an upgrade that will represent an enhanced asset in the developer's portfolio and enable it to compete with an average new-build product in the local market. Buildings that are multi-occupied are often suited to this approach, allowing landlords to provide a major change every ten to fifteen years.

Level 4 – Comprehensive Refurbishment

This option is the most expensive of the refurbishment options and carries the highest development risk. However, it creates the best opportunity to capitalise on the improvement in asset value and associated increases in rent and aims to attract a wider base of potential tenants. This level includes the works outlined in the previous levels plus further development opportunities outside the building. For a site that has a particularly high residual land value, major refurbishment options can be more financially viable than demolition and new build. The works will bring the building up to current standards and 'future-proof' it for 20–25 years. Level 4 considers fabric performance and the lifespan of materials as well as the running and maintenance costs of the fully occupied building. Issues such as the relocation of the complete plantroom to optimise floor space, by using the previous plantroom areas, and the introduction of more efficient plant machinery at roof level or new plant towers to the sides of building can be considered, subject to structural and planning limitations.

This level of refurbishment intervention and the associated levels of investment can extend the lifespan of a building by bringing all elements up to date and ensure the building is competitive with high-value new-build office accommodation in the local market. The whole building will be affected, so this option is most appropriate for an empty building or one at the end of its lease. The ability to extend the building and add floors is often considered at this level of refurbishment. In addition, development on land associated with the building, such as air-rights development above surface car parking, can be considered. This will enhance the value of the site and help deliver more area, potentially a wider range and mix of uses, and increased environmental credibility. At the extreme end of 'Comprehensive Refurbishment', only the structure might be retained, with complete replacement of the exterior envelope, services, cores etc.

Level 5 – Demolition

The final level covers the demolition of the whole building, which will enable the construction of a new facility or amenity space in the vacant area. The choice of demolition will have sustainable imperatives associated with it and also provides a baseline in terms of cost for a new building facility in its place.

Table 1.2, below, provides the costs associated with the refurbishment options calculated as an average figure, but also with the lower and upper quartile costs. As expected, the lower the degree of refurbishment, the lower the cost. However, at higher levels of refurbishment, the difference in costs between the upper value of one level and the average level of the next level is minimal and sometimes even lower. Understanding the risks for greater levels of refurbishment is critical to defining the costs, with errors of margin likely to be significant in determining the optimum solution for the building.

Different buildings will require varying levels of refurbishment treatments dependent on the occupants and use of the building. Typically, multi-tenanted buildings will undergo light-touch refurbishments to core areas at the time when tenants change to maintain a fresh face to the building, supported by an extensive refurbishment every 15 years to ensure the building is updated both in plant areas and across the tenanted areas. Similarly, an owner-occupied building typically will undergo a medium refurbishment either across the whole building or in parts.

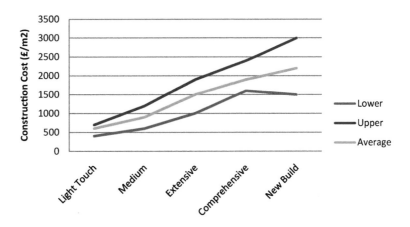

Figure 1.1: Cost assessment of refurbishment versus new-build construction costs.

Table 1.2: Cost assessment of refurbishment versus new-build construction costs

Degree of Refurbishment	Lower		Upper		Average		Average Difference from New Build
	£/m2	£/ft2	£/m2	£/ft2	£/m2	£/ft2	
Light Touch	400	37	700	65	600	56	−73%
Medium	600	56	1200	111	900	74	−59%
Extensive	1000	93	1900	177	1500	139	−32%
Comprehensive	1600	140	2400	223	1900	177	−14%
New Build	1500	139	3000	279	2200	204	−0%

In addition, there are the associated time reductions to bring the refurbished building to market as described in Figure 1.2.

Whilst refurbishment covers a range of diverse activities, the procurement options available do not generally cater specifically for the particular characteristics of refurbishment projects. In developing a strategy, the starting point is to understand the complexity of the project. The ability to influence and drive sustainable measures is also different across the various levels, with Levels 1 and 2 more limited by virtue of the reduced scope of work.

Key times for intervention for refurbishing include:

- When a significant gap emerges between the rents being achieved in your property and those in the same or equivalent locations;
- Your building loses a major tenant or multiple tenants and there are prolonged periods of vacancy;
- A major tenant's lease is approaching its end and refurbishment offers an incentive to stay;
- Major plant requires refurbishing.

Simple refurbishment projects such as those covered in the Light Touch might be undertaken as part of a planned refresh cycle, or might be a short-term tactical investment to extend the economic life of an asset. The timeframe of the investment is typically five to seven years. There are a few complexities associated with

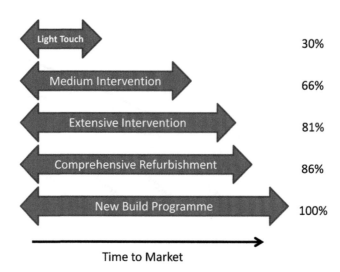

Figure 1.2: Time to market for the various levels of refurbishment.

this level of refurbishment such as elements of building services needing to be updated.

Medium-level projects generally involve upgrades to building services so the frequency investment is on a cycle of 15 to 25 years. Increasingly, improvements to the building fabric are required as a consequential requirement to meet Building Regulations. Refurbishments of this kind involve a greater level of risk associated with the existing building fabric and systems – either related to the reuse of some systems or the replacement or remodelling of windows and risers. The condition of the existing fabric and systems, together with any effects for end-users if refurbishment takes place within an occupied building, will have an impact on the level of risk for the project.

Major refurbishment is aimed at long-term remodelling of a building, addressing constraints such as circulation and maximising the potential offered by the site and the building consent. The risk with major projects is much greater but it can be met by the ability of an appropriate contractor to manage the risks. The maximum amount of information on building condition and any other risks need to be identified in advance.

1.3   Building Refurbishment Market and Size

In most developed countries, the existing building stock accounts for over 98 per cent of the total stock, with a replenishment rate of up to 2 per cent in peak times and lower than 1 per cent during times of economic difficulties[5]. Many of these structures were constructed prior to the recent global improvements in Building Regulations and Energy Standards. In the UK, more than 77 per cent of the commercial building stock was constructed before the energy-conservation measures were enforced[6]. This level is replicated across many other developed countries.

Conversely, the level of new build in the high-growth developing countries is significantly higher as new infrastructure is built to provide workplaces and homes for many of those arriving in the rapidly growing urban centres. India and China,

Figure 1.3: Existing built structures in Spain (a) and India (b) constructed to historic building standards requiring different levels of refurbishment.

in particular, have seen a significant increase born from such an urbanisation together with the increase in manufacturing and employment capacity to meet the requirements of global trade. Mostly being removed are poorly constructed buildings in order to increase the urban space and provide more stable and substantive building. As such, the level of refurbishment is more likely to be Level 5 (Demolition) than any other (Figure 1.3).

The refurbishment market is also affected globally by economic changes and investment effects which drive the use and availability of space together with the ability to provide the capital for upgrades to buildings. It is interesting to note that cities in developed countries have a significant refurbishment market due to lack of space in urban centres, which helps to support the refurbishment market. Conversely, the new-build programme of many developing countries with expanding cities or new urban centres means a limited focus on refurbishing existing buildings.

Analysis by MBD shows that the repair-and-maintenance market is approximately 45% to 50% of the total building construction sector[7] within the UK amounting to a spend in excess of £50 billion per annum. This figure includes a range of sectors and scale of refurbishment activities including commercial and domestic improvements. The report states: 'In the private non-residential sector, repair & maintenance expenditure has been increased strongly within the review period, reflecting a reduced level of new construction expenditure within the industrial sector in the early part of the review period as well as a significant slowdown in new construction activity within the commercial sector. Furthermore, although there has been economic growth, corporate profitability has remained under pressure encouraging repair & maintenance over new investment.'

Adaptive Re-use of Buildings – Portland, Oregon – Case Study

Where tight capital and weak markets for greenfield projects exist, adaptive use is an attractive approach to developers and to governments looking to create local jobs and fight climate change. Adaptive-use projects, aligned with inner-city regeneration, remain a bright spot and are commonly supported by government incentives.

In October 2009, President Obama issued an executive order that required the federal government to redevelop its own massive inventory of buildings through adaptive use and green renovation. Under the order, planning for new facilities or leases also has to consider sites that are pedestrian-friendly, accessible to public transit, and within existing central cities. As an example, the Edith Green–Wendell Wyatt Federal Building in downtown Portland, Oregon will receive $130 million in federal stimulus money for a state-of-the-art sustainability retro-fit. The project will include a rooftop solar array, 50 per cent reduction in lighting energy, 65 per cent reduction in water use, and a renovated energy-efficient exterior.

However, it takes a significant set of additional skills to manage the challenges of adaptive use, involving a complex mix of resources: federal historic tax credits, federal brownfield development loans, state historic tax credits, state brownfield credits, tax increment financing, property tax abatements and other tools. Partners have often included governments, institutions and other businesses that have to find new structures to work together effectively. The key is to unlock the hidden economic benefits of adaptive use, in combination with other forms of sustainable development, so that the projects become fully market-driven. That requires finding and combining all potential revenue streams.

The new emphasis on adaptive use is coming partly from sustainability and economic development goals. A bigger factor is a tidal wave of demographic change, combined with changing spending habits. ULI's *Emerging Trends in Real Estate 2011* documents the trend: 'Infill over fringe – the ''move back in'' trend gains force as 20-something echo boomers want to experience more vibrant urban areas and aging baby boomer parents look for greater convenience in downscaled lifestyles.' Those demographic forces, together with changing markets and continued tight credit, mean a seismic, perhaps permanent, change in the development environment. This is providing a real focus on 'repurposing' existing buildings over the next three years, especially with the need to focus on urban areas.

Even with these incentives, there remain challenges for adaptive use and renovation, particularly where these are uncovered late in adaptive-use projects. Building codes act as an impediment, by not enabling a historical mixed-use project that has been reduced to a single use to be reconverted. Governments seem increasingly motivated to remove these barriers, in part because of the job-creating potential and other economic benefits of adaptive-use projects. Advocates point to the historical successes of projects like Vancouver's Granville Island, an arts-and-market district where an initial public investment of just CDN$25 million now produces CDN$35 million in annual tax revenue alone – an eye-popping return of 140 per cent. Add to that an estimated 2,500 new jobs, and CDN$130 million a year in additional economic activity.

While those kinds of returns are not often duplicated, research shows that such local economic activity does bring a bigger economic-multiplier effect for the community. That effect can be especially powerful with adaptive use of historical buildings.

The US building stock is constantly in flux, but presents a huge opportunity for change. Each year approximately 1.75 billion square feet of the nation's 300 billion square feet of building space are demolished and replaced with approximately five billion square feet of new building space. In addition, about five billion square feet of building space are remodelled each year. Commercial buildings in particular are viewed as the best targets for improvement. In the US, for example, there are 74 billion square feet of non-residential space.

There was a prolific period of commercial-office development from the 1970s to the 1990s when many buildings constructed were designed to last 25 to 30 years and are therefore now reaching the end of their life and coming to the market as redevelopment opportunities. Importantly during this timeframe, the buildings were constructed to meet a speculative market and therefore constructed with a lower cost in mind to maximise profits.

Historically, buildings were constructed for occupation by multiple users who would be able to adapt and modify the interiors, and exteriors, to suit the requirements of the market. Larger houses became offices or hotels and were adapted to suit. However, there has been a trend more recently for buildings to be constructed for a finite use, reducing the ability for the building to be refurbished for alternative uses.

This has led to a situation whereby the refurbishment market is focused towards older building stock where it is feasible to make the changes. More modern buildings will incur a significant cost to bring them back to market and therefore it becomes cheaper to demolish and rebuild.

This is forcing decisions to be made about the investment benefits of refurbishing the building relative to demolishing it and starting afresh. Buildings built between the 1970s and the 1990s are often defined by certain technical criteria and dimensional characteristics and have been impacted by incredible technological evolution over that period.

Growth in developing economies is still the priority, which leaves little potential for refurbishment on a major scale as it is easier to knock down and rebuild. The key for these economies such as in India and China is to learn from experiences within the US, Europe and similar countries and adopt green building practices. Nevertheless, some economies operating within constrained space requirement are looking to better utilise the buildings they have. Since 2005, Malaysia has been moving away from constructing new buildings in favour of refurbishing historic and old ones. This is due to a number of reasons, including the economic crisis, land limitation and sustainable issues.[8]

Endnotes

1 Webster's New World College Dictionary Copyright © 2009 by Wiley Publishing, Inc., Cleveland, Ohio. Used by arrangement with John Wiley & Sons, Inc.
2 LEED New Construction V2.2 Reference Guide October 2007
3 BREEAM user guide – www.breeam.org/
4 Can Do Refurbishment: Commercial Buildings of the 70s, 80s and 90s, British Council for Offices, October 2009
5 I. McAllister and C. Sweett, 'Transforming Existing Buildings: The Green Challenge', March 2007
6 'Surveyors lead plan to transform existing commercial buildings', Energy in Buildings and Industry, 2007; vol. 4, p. 8
7 The UK Facilities Management Market Development © MBD 2007
8 Occupant feedback on indoor environmental quality in refurbished historic buildings; S. N. Kamaruzzaman, M. A. Emma Zawawi, Michael Pitt and Zuraidah Mohd Don; International Journal of Physical Sciences Vol. 5(3), pp. 192–199, March 2010, www.academicjournals.org/IJPS

2 Sustainable Refurbishment

It is easy to become interested in all the exciting new-build schemes. They look good, they test innovation and they are good publicity. But the biggest environmental, social and economic impacts of the built environment act through the existing stock.

Of the stock that will exist in 2030 in developed countries, 70 per cent is already in existence: new build only adds 1–2 per cent to the built stock each year within the UK. The rest is likely to have been refurbished – possibly several times – and the nature of these refurbishments is vital for the health and productivity of people who live and work in these buildings, as well as for reducing our environmental impacts.

The challenge for refurbishing and upgrading existing stock is immense – refurbishment projects obviously need to upgrade the structure, the fabric and the building's services whilst complying with new standards and legislation, but they should also:

- Address the potential effects of climate change;
- Have minimal environmental impact;
- Conserve our heritage buildings;
- Provide more safe and secure internal environments;
- Produce spaces that are adaptable for change of use;
- Demonstrate best value though procurement and partnering;
- Contribute to regenerating local communities.

Sustainable Refurbishment, First Edition. Sunil Shah.
© 2012 Sunil Shah. Published 2012 by Blackwell Publishing Ltd.

Chapter Learning Guide

This chapter will provide an overview of sustainability in respect of building refurbishments, including:

- Overview of sustainable development and the key themes that are discussed under its umbrella;
- Review of cross-cutting areas of climate change affecting building requirements;
- Use of the 14 themes of sustainability to define the typical areas of sustainable refurbishment;
- Global Alliance for Building Sustainably (GABS) Charter.

Key messages include:

- Sustainable development is a constantly changing area comprising a number of key themes;
- Sustainable refurbishment is developing as an area, but there are relatively few modern examples.

2.1   Introduction

Sustainable refurbishment looks at our built heritage in a holistic manner – taking into account not only the physical condition of a building and its energy performance, but also the impact of the building on the community as a whole. It enhances, conserves and promotes our existing built heritage for present and future generations. In environmental terms, many older buildings (particularly eighteenth-century and nineteenth-century buildings) have key characteristics that make them more adaptive and resilient: modest plan depths, high thermal mass, high ceilings and narrow windows. There are lessons to be learnt from the built heritage that has survived thus far. Heritage features are valued by occupiers, but have to be part of a functional whole.

Benefits of Sustainable Refurbishment

- *Preservation of built heritage:* Sustainable refurbishment protects our existing built environment for current and future generations.
- *Reduced environmental footprint:* Incorporating high standards for energy efficiency, it generates a reduction in greenhouse-gas emissions to the environment, reducing the threat from climate change. Higher standards are also extended to water systems and lighting. Installation is managed to ensure the use of fewer polluting materials, efficient waste management and incorporation of recycling capabilities.

> - *Better adaptation to climate change:* Designed to limit the effects of solar gain in summer, improving water efficiency, better management of surface-water run-off.
> - *Lower running costs:* Greater energy and water efficiency helping to reduce poverty.
> - *Improved well-being:* More pleasant and healthy place to live, with natural ventilation and lighting.
> - *A mark of quality:* Independent accreditation means the building can be differentiated from those of competitors.

2.2   Overview of Sustainable Development

Whilst the term, sustainable development, has gained popular momentum over the last ten years, the components of it date back many decades. From *Silent Spring* written by Rachel Carson[1] in the early 1960s, describing a world affected by chemicals and the drive for increased productivity, through to James Lovelock's Gaia philosophy[2] stating the role of 'mother earth', sustainable development has formed a spine through these books.

One of the first definitions of sustainable development was made in 'Our Common Future', the report of the Brundtland Commission, calling for development **'that meets the needs of the present without compromising the ability of future generations to meet their own needs'**[3]. Whilst this definition is still used today, over 500 definitions of sustainability and sustainable development have been spawned by various governments, professional bodies, institutions and organisations. A more commonly known terminology encompasses the environmental, social and economic principles captured as the 'triple bottom line'.

The UN Global Compact[4] has three principles devoted to the environment, demonstrating how important the environment has become: the precautionary, proximity and polluter pays principles. This effectively means the greater the environmental impacts of a material, the greater the cost to deal with it – from production, use and disposal. The onus is therefore on trying to avoid producing the material in the first instance by looking for alternatives. As far as the environment is concerned, it is not just a question of understanding cause and effect, but also the relationships between the different species. It is also recognition of the fact that some changes, once they occur, are irreversible.

Sustainable development has been increasingly quoted over recent years as more organisations jump on the 'triple-bottom-line' bandwagon. The term has a variety of meanings dependent on the requirements of the organisations. This is largely as a result of a lack of guidance provided to fully define and capture what sustainable development means for specific sectors and activities. Whilst the provision of social, environmental and economic partnership is well provided for, the combinations and elements of the partnership are less well understood. This has meant a relatively unstructured process for organisations moving towards implementing sustainable-development activities and procedures into their day-to-day operations.

The current approach towards sustainable development is captured in Figure 2.1, which shows the environment, social and economic principles in increasing

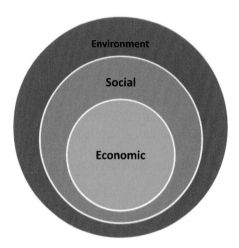

Figure 2.1: Approach to sustainable development.

circles similar to the concept of Russian dolls. The outer circle is the environment – we have one planet that we occupy and inhabit with a finite set of resources that we can use. There are no means to generate more materials than those already present and so we must prioritise their use. The next circle within that of the environment is social – society lives within the environment we have and must operate within its confines and the resources that are provided. The final circle within both the environment and social is economy – our consumption of materials and goods is based upon the finite resources and our ability to utilise the materials in the optimum fashion.

The following part provides brief details of the various themes that are typically covered under the heading of sustainable development. The list is by no means comprehensive, but does intend to cover the key areas that will affect the majority of organisations.

Population Growth

The future population growth of the world is difficult to predict, with the UN Census Bureau and the US Census Bureau giving different estimates. According to the latter, world population will hit seven billion in July 2012[5], or by late 2011, according to UN prediction[6]. Birth rates are declining slightly on average, but vary greatly between developed countries (where birth rates are often at, or below, replacement levels), developing countries, and different ethnicities. Death rates can change unexpectedly due to disease, wars and catastrophes, or advances in medicine.

Figure 2.2 shows the current population trends from 1950 to the present day and modelling on a variety of scenarios to 2050[7]. The current trajectory is on the medium-to-high scenario providing a population in the region of 8.5 billion to 10.5 billion people by 2050. Such a rise would provide a quadrupling of the global population in the past 100 years, placing a massive burden on the planet for resource needs together with dealing with the emissions and pollution. This is occurring at the same time as that of our level of material consumption increasing dramatically.

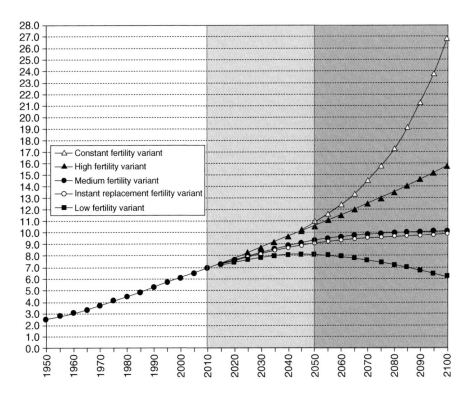

Figure 2.2: Estimated and projected world population according to different variants, 1950 to 2100 (billions). Source: United Nations, Department of economic and Social Affairs, Population Division (2011): World Population Prospects: The 2010 Revision, New York (updated 18 April 2011).

The Worldwatch Institute has highlighted that more than a quarter of the world's population has now entered the 'consumer class'. Higher levels of obesity and personal debt, chronic time shortages, and a degraded environment are all signs that excessive consumption is diminishing the quality of life for many people. Private-consumption expenditure at the household level has increased fourfold since 1960. The 12 per cent of the world's people living in North America and Western Europe account for 60 per cent of this consumption, whilst the one-third living in South Asia and sub-Saharan Africa account for only 3.2 per cent[8].

The report points out that in the US today:

- There are more private vehicles on the road than people licensed to drive them;
- The average size of refrigerators in households increased by 10 per cent between 1972 and 2001 and the number per home rose as well;
- New houses were 38 per cent bigger in 2000 than in 1975, despite having fewer people in each household on average.

Climate Change

The world's climate is changing and will change further in the coming decades as a result of an increasing concentration of greenhouse gases (GHGs), notably carbon dioxide, which accounts for 60 per cent of the total impact of GHG emissions. Since the Industrial Revolution, levels of carbon dioxide in the atmosphere have grown by more than 30 per cent as a result of burning fossil fuels, land-use change and other man-made emissions. This is amplifying the natural 'greenhouse effect', leading to global warming: the average surface temperature increased by 0.6 degrees Celsius during the twentieth century. The 2007 report of the Intergovernmental Panel on Climate Change (IPCC)[9] concluded that 'Warming of the climate system is unequivocal, as is now evident from observations of increases in global average air and ocean temperatures, widespread melting of snow and ice and rising global average sea level' (see Figure 2.2). Most of the observed increase in global average temperatures since the mid-twentieth century is very likely due to the observed increase in anthropogenic GHG concentrations.

The period from the 1980s onwards has been estimated to be the warmest period in the last 2000 years, and 11 of the 12 hottest years on record occurred in the period from 1995 to the latest records in 2006. The IPCC estimates that global average surface temperatures could rise by a further 1.1 to 5.4 degrees Celsius by 2100. Rising emissions and a disrupted climate are leading to a range of impacts, such as more frequent heatwaves, increased intensity of floods and droughts, as well as rising sea levels.

These changes are already translating into real economic losses:

- In 2002, the severe floods across Europe generated direct losses of $16 billion[10]. Further severe flooding has taken place across central Europe in 2005, 2006, 2009 and 2010 with associated significant insurance costs;
- The 2003 heatwave that affected much of Europe is estimated to have caused 35,000 premature deaths and had an estimated economic cost of $13.5 billion[11];
- Hurricanes Katrina and Wilma in 2005 battered the Gulf Coast in the United States causing widespread damage and loss of life;
- Claims for storm and flood damages in the UK tripled over the decade to 2009 (£4.5 billion) compared with the previous decade (£1.5 billion), according to the Association of British Insurers[12];
- The Stern Review calculates that the cost from dangers of unabated climate change would be equivalent to at least 5 per cent of global gross domestic product (GDP) each year compared with a cost of 2 per cent GDP to achieve stabilisation[13].

Climate change is a natural phenomenon and the Earth does vary in temperature based upon a number of cyclical activities. As such, there are natural changes in the temperature including global-warming periods and global-cooling situations that have led to previous ice ages. These are natural fluctuations in temperature and carbon dioxide levels, evidenced through investigation of ice cores from the Poles.

It is widely believed that the gases emitted as we burn fossil fuels are the most likely reason for the recent period of warming during the past century. These gases contribute to the 'greenhouse effect' as they accumulate in the atmosphere, trapping the outgoing heat radiated from the surface of the Earth. Currently the planet is increasing in temperature over the past several thousand years since the last ice age on a natural increase in temperature. Natural variability of climate due to solar output and volcanic eruptions partly explains the recent warming. However, the role of mankind is exacerbating the natural variation in temperature through

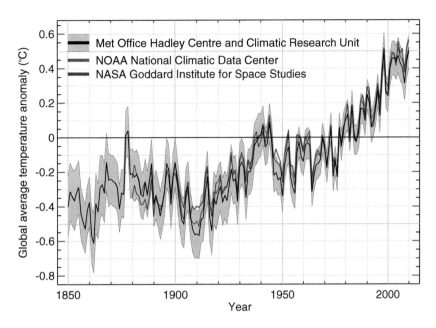

Figure 2.3: The three main records of global average surface temperature. Red line = NOAA record, blue line = NASA record, black line = Met Office/ UEA CRU record, with grey shading showing 95% confidence interval on Met Office/UEA CRU record. Source: Met Office Hadley Centre. Before 1850, instrumental time series measurements with global coverage are not available. [IPCC AR4 (2007) (Working Group 1; 1.3.2)].
© Crown Copyright 2011, Met Office.

the addition of contaminants into the atmosphere faster than the planet can 'manage' itself and normalise the effects (see Figure 2.3)[14]. This is leading to a faster warming phase than would otherwise be expected on a global basis, which does mean that some areas will warm more quickly and sharply than others.

The effects of climate change are going to be a greater variability in weather patterns and an increase in major environmental events. Climate change will bring about changes in weather patterns, rising sea levels and increased frequency and intensity of extreme weather leading to rising temperatures and greater risks of flooding. Additional changes will result in droughts, floods, erosion and damage to buildings and roads[15].

Use of Resources

The increasing consumer demand is placing an increasing burden on the world's natural resources and habitats. The Worldwatch Institute has classified the 'consumer class', once limited to the rich nations of Europe and the US and epitomised by large cars, plentiful diets and an abundance of waste, as now encompassing almost a quarter of humanity. Since 1960, the amount spent on household goods and services has increased fourfold, exceeding US$20 trillion in 2000. However, the developing world is beginning to copy the trends of the richer nations, with China and India now home to a larger consumer class than that in all of Western Europe.

The rising consumption in the US, other rich nations and many developing ones is more than the planet can bear. Forests, wetlands and other natural places are shrinking to make way for people and their homes, farms, shopping malls and factories. There is also a growing scarcity of key materials that we require, including copper and lead, pushing up prices for their procurement but also limiting their availability to many nations.

World Living Beyond Its Environmental Means – WWF[16]

In 2007, the most recent year for which data are available, the footprint exceeded the Earth's biocapacity – the area actually available to produce renewable resources and absorb carbon dioxide – by 50 per cent. Overall, humanity's ecological footprint has doubled since 1966. This growth in ecological overshoot is largely attributable to the carbon footprint, which has increased elevenfold since 1961 and by just over one-third since 1998. However, not everybody has an equal footprint and there are enormous differences between countries, particularly those at different economic levels and levels of development.

The latest ecological footprint shows that this trend is unabated (Figure 2.4). In 2007, humanity's footprint was 18 billion gha, or 2.7 gha per person. However, the Earth's biocapacity was only 11.9 billion gha, or 1.8 gha per person. This represents an ecological overshoot of 50 per cent. This means that people used the equivalent capacity of 1.5 planets in 2007 to support their activities.

What does overshoot really mean?
How can humanity be using the capacity of 1.5 Earths, when there is only one? Just as it is easy to withdraw more money from a bank account than the interest this money generates, it is possible to harvest renewable resources faster than they are being generated. More wood can be taken from a forest each year than regrows, and more fish can be harvested than are replenished each year. But doing so is only possible for a limited time, as the resource will eventually be depleted. Similarly, carbon dioxide emissions can exceed the rate at which forests and other ecosystems are able to absorb them, meaning additional Earths would be required to fully sequester these emissions.

The ecological footprint shown in Figure 2.4 describes the increasing number of planets required to meet the human population's resource requirements. Total biocapacity, represented by the dashed line, always equals one planet Earth, although the biological productivity of the planet changes each year. The footprint of traditional activities, including grazing and fishing, has increased by some 50 per cent over the past 45 years. However, the overextension of human-resource requirements is largely due to the substantial increase in carbon emissions.

Ecological footprinting has been developed as a mechanism to translate the use of resources and materials to enable easy comparison. The globe has an

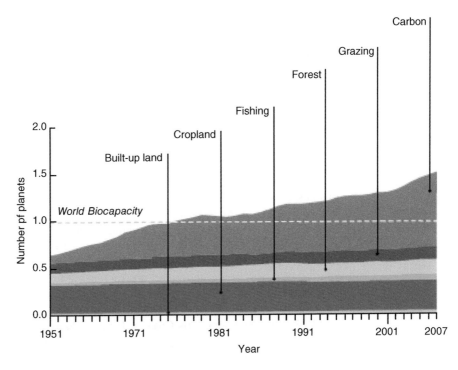

Figure 2.4: Ecological footprint by component, 1961–2007 (Source: WWF/ZSL/GFN).

average footprint of 1.8 hectares per person. Each activity, whether at country, city or business level, has an ecological footprint that can be compared to this average. Examples of ecological footprints from the Global Footprint Network[17] include:

United Arab Emirates	10.7 ha/person
United States	8.0 ha/person
Australia	6.8 ha/person
New Zealand	4.9 ha/person
United Kingdom	4.9 ha/person
Russia	4.4 ha/person
Brazil	2.9 ha/person
China	2.2 ha/person
India	0.9 ha/person

Fossil fuels are being depleted 100,000 times faster than they are formed; each person in the Western economies consumes 60 to 80 tonnes of material each year on average, and it requires roughly 11 tonnes of raw materials to produce one tonne of product.

The world's forests are still being destroyed mainly from the conversion of forests to agricultural land. The numbers measure net loss, taking into account forest growth from new planting and natural expansion. An average 7.3 million hectares was lost annually over the last five years. This was down from 8.9 million hectares (22 million acres) a year between 1990 and 2000[18].

What does this mean? Ultimately, the overuse of resources will lead to a shortage of materials, resulting in either a move towards alternatives, to resource efficiency, or to a reduction in consumption. It is likely that more than one of these events will take place for many materials, largely due to the cost of supply.

Waste

The common view of waste is the 'end of pipe' approach to materials that have no further use and are therefore thrown away. In many developed nations, historically this has meant disposal to landfill but more recently has included the recycling and reuse of such materials.

There are major differences globally in landfill levels. In 2006, approximately 76 per cent of municipal waste in England was sent to landfill sites, compared with around 40 per cent in France and 0 per cent in Switzerland and Germany[19]. The volumes of waste being generated are increasing from the rising amount of packaging materials and consumer-led changes in products such as electronic goods.

The EU-27 Member States plus Croatia, Iceland, Norway and Turkey in total generated some 3 billion tonnes of waste in 2006, or roughly 6 tonnes per person, of which around 3 per cent is hazardous (Eurostat data centre on waste, 2010; data reported according to the Waste Statistics Regulation). In general, 32 per cent of the waste generated in the EEA countries is from construction and demolition activities, 25 per cent from mining and quarrying, and the rest from manufacturing, households and other activities. About two-thirds of the total is mineral waste, mainly from mining, quarrying, construction and demolition. On a global level, North America produces almost double the municipal waste of that of Europe, and almost four times of that of Asia[20] (see Figure 2.5).

Factors underpinning poor environmental performance on waste include the ready availability of cheap landfill sites; weaker regulatory controls and the absence of incentives for recycling; low public awareness; and an inability or unwillingness on the part of many local authorities to invest in more expensive recycling and waste-disposal options. Examples of general disposal and general poor management of waste are commonplace across the globe. In order to reduce the volume of waste generated, radical action is required. Where a fixed rate for waste management is paid, there is no economic incentive to reduce waste volumes, or to recycle and compost.

A variable rate for waste management encourages recycling so that good recyclers are rewarded whilst those who produce more rubbish have to pick up the bill for their wastefulness. Such schemes are commonplace in countries with high recycling rates (for example, Germany and Austria) where they have typically led to an increase of 30 per cent to 40 per cent in recycling and composting.

Recycling also has a number of adverse environmental impacts. For example, waste has to be collected and transported to recycling facilities, which might be further away than landfill sites. The waste then has to be processed back into useful materials, and the relevant processes might be relatively energy-intensive and/or polluting. The only way to arrive at a clear answer as to whether recycling is environmentally beneficial is through a process called life-cycle analysis (LCA),

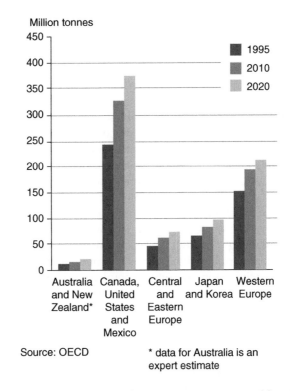

Figure 2.5: Projected trends in municipal waste generation in world regions.

which seeks to measure all the environmental impacts of resource use, for both virgin and recycled materials, so that they can be validly compared.

Using the LCA methodology, it is clear for some materials, especially metals like aluminium and steel, that recycling has environmental benefits and is often cost-effective too, because the energy required for recycling is much less than for producing these materials out of virgin ore.

For glass, the energy balance tends also to be positive, but there can be problems with the product, because recycling green glass, which dominates the waste-glass stream, will not produce clear glass. Some of the benefits of recycling can be lost if the recycled glass is put to a different use (for example, as aggregates in roads). This is still preferable to using raw materials as it reduces the amount of waste being landfilled.

For paper and cardboard, the environmental issues are more finely balanced: they depend on the distances the waste has to be transported, on how the trees for virgin paper are grown, and whether they are replanted once cut down. Sometimes it is clear that the best use for waste paper and cardboard in terms of energy would be for energy generation in incinerators, but this raises other environmental issues such as air emissions.

Surveys show that recycling is one of relatively few environmental actions that command widespread public support, probably because it is one of relatively few instances where individual action can be seen to make a difference[21]. It could be counter-productive to reduce people's environmental commitment in respect of recycling and perhaps impacting on other issues, by drawing attention to the complexities of recycling's balance of benefits, or, even worse, to those relatively rare cases where recycling is not environmentally beneficial.

Pre-treatment of waste as a means to reduce the volume of waste to landfill is becoming enshrined within legislation and corporate practice. The removal of hazardous materials such as printed-circuit boards from electronic equipment and contaminated materials enables the bulk of the waste to become recycled. Much of the pre-treatment takes place in parts of the world where labour is cheaper and therefore where poor health-and-safety practices and low wages predominate, such as in Gujarat State in India.

Discarded refrigerators are exported to developing countries, burnt in incinerators or find their way onto landfill sites. The majority of refrigerators, built prior to 1996, contain harmful ozone-depleting substances (ODS). In the EU, electronic waste is the fastest-growing waste stream, with the average EU citizen producing 17–20 kg of electrical waste every year, much of which contains potential pollutants. The vast majority of the waste in Europe still goes to landfills or incineration, despite general acceptance of the need for action to reduce inefficient use of resources and the risk of contaminants leaking into the surrounding soil, water or air, posing a risk to human health as well as the wider environment.

Aside from environmental issues, there are major commercial advantages to recycling, including the recovery of valuable metals such as iron, steel, copper and aluminium from white goods, such as refrigerators, which typically consist of 70 per cent metal. Once recovered, the metals can be sold back to the metals' industries for recycling, to generate revenue. This produces not only economic benefits, but also environmental benefits. Recycling metals uses less energy. For instance, 20 times the amount of energy to produce one tonne of recycled aluminium is required to yield one tonne of new aluminium. As a consequence, the need to locate and mine new metal ores is reduced, placing less pressure on the environment.

Water

The earth is covered by 70 per cent water, yet there is a known scarcity of water available in the world. Flooding affects Bangladesh and the south coast of the US on a regular basis, with water shortages causing conflicts between Egypt, Ethiopia and Sudan from the river Nile, and in the Middle East from the river Euphrates. Municipal water-consumption patterns represent the availability of water: the US consumes 215 cubic metres per person per day; India consumes 47.4; China 50.53; and Bahrain 234.1[22].

Of the global provision of water, only 2.5 per cent is fresh water available for human use, with much of this locked up in ice caps and glaciers. The total accessible water is only 0.0125 per cent of the global total, of which 1.1 billion people do not have sufficient access to fresh water. As the world population continues to grow, the accessibility of sufficient water is becoming a significant issue together with water stress as current levels of water supply become overstretched to supply the growing population.

Pollution of existing water systems leads to diseases that can kill people and it also limits supplies, hindering development (Figure 2.6). In 2002, 2.6 billion

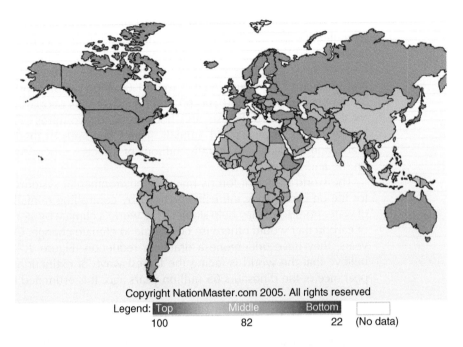

Legend: Top   Middle   Bottom   (No data)
100   82   22

Figure 2.6: Regional and global trends in the access to clean drinking water.

people were without improved drinking water – over 40 per cent of the global population[23].

- Five litres of oil poured onto a lake can spread to cover an area the size of two football pitches;
- Most pollution incidents are the result of ignorance, apathy or neglect;
- Just one litre of solvent is enough to contaminate 100 million litres of drinking water.

Biodiversity

There are many issues affecting biodiversity – but perhaps the two greatest issues are the habitat loss from forest destruction and oceans. Recent drives to improve the level of biodiversity, such as small-scale forestry, has resulted in non-indigenous planting, causing greater damage through the introduction of species unnatural to the local food chain.

Ten million hectares of ancient forest are being cleared or destroyed every year – the equivalent to an area the size of a football pitch every two seconds will have disappeared. Whilst there are around 1,350 million hectares of ancient forest remaining undisturbed, this represents only seven per cent of the Earth's land surface, and only one-fifth of the forests' original size. Since 1950, 20 per cent of the world's ancient forests have been cleared, with those in Indonesia and Central Africa likely to have gone in a few decades if forest destruction continues at its present pace[24].

The primary causes of forest loss and degradation vary, but revolve around agricultural expansion, mining, settlement, shifting agriculture, plantation establishment and infrastructure development. These ancient forests are home to millions of forest people who depend on them for their survival – both physically and spiritually. It is estimated that some 1.6 billion people worldwide depend on forests for their livelihood and 60 million indigenous peoples depend on forests for their subsistence. Forests also house around two-thirds of the world's land-based species of plants and animals. That's hundreds of thousands of different plants and animals, and literally millions of insects – whose futures also depend on the ancient forests[25].

The world's ancient forests maintain environmental systems that are essential for life on Earth. They influence weather by controlling rainfall and evaporation of water from soil. They help stabilise the world's climate by storing large amounts of carbon that would otherwise contribute to climate change. Over the past 8000 years, they have undergone a dramatic reduction (Figure 2.7). Many scientists believe that the world is facing the largest wave of extinctions since the disappearance of the dinosaurs 65 million years ago. It is estimated that nearly 24 per

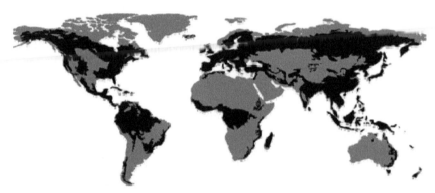

Ancient forests 8000 years ago...

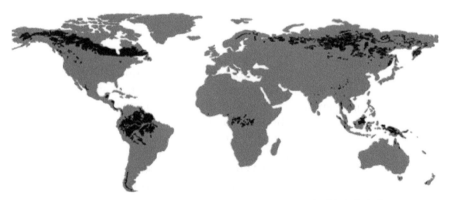

ancient forests today

Figure 2.7: Disappearance of native ancient forests over the past 8000 years (Source: World Resources Institute, Washington 1997, revised by Greenpeace 2002).

cent of mammals, 12 per cent of birds and almost 14 per cent of plants face extinction. Most of these extinctions will be due to habitat destruction; most of these habitats are ancient forests.

Recent research by the World Resources Institute (WRI) concludes that 'commercial logging poses by far the greatest danger to frontier forests . . . affecting more than 70 per cent of the world's threatened frontiers'[26]. The way the logging industry operates also exacerbates the problem. Unplanned tree cutting and inefficient processing leads to an enormous wastage of wood, whilst lack of transparency within the industry makes it very difficult to trace the exact source of wood supply. Illegal logging is adding to this problem accounting for up to 90 per cent of the logging that takes place in Indonesia, 60–80 per cent in the Brazilian Amazon and 50 per cent in Cameroon between 1999 and 2004.

The oceans cover two-thirds of the planet, host 80 per cent of life from microscopic plankton to great whales and provide half the oxygen the planet needs. The increasing exploitation of marine life together with the burning of fossil fuels and the development and dumping of chemicals into the oceans outpace each ocean's ability to cleanse itself and maintain a natural balance[27]. Commercial fishing and tourism are placing a strain on the oceans to replenish themselves:

- Industrial-fishing fleets are far exceeding the oceans' ability to recover, with smaller fish sought as larger fish are wiped out. It is estimated that 90 per cent of the world's large fish are already gone;
- Every year, up to 300,000 whales, dolphins and porpoises die in nets and 100,000 albatrosses are caught on hooked fishing lines. Turtles, seals and sharks are also victims of indiscriminate fishing practices;
- The extension of fishing grounds into the waters of Africa and the Pacific is increasing the conflict between countries over fishing rights;
- In 1998, 16 per cent of the world's corals were severely damaged and in South Asia and the Indian Ocean, half of the reefs lost the majority of their coral;
- Chemicals, oil, plastics, sewage, industrial discharges, intentional dumping, mining, general rubbish and tanker incidents are causing severe marine pollution.

2.3   Sustainable Development and Building Refurbishment

Sustainable development is becoming increasingly important to organisations through the development and implementation of sustainable policies, which translate the strategy into action and demonstrable evidence of progress. Increasing numbers of companies throughout the building-supply chain, including investors, property owners and service-delivery companies, are developing and implementing policies. The unresolved challenge is to coordinate the policies and activities of the parties in this chain so that each can demonstrate that its policy is being observed and its objectives met.

Refurbishing buildings in one sense can be said to be inherently sustainable, helping to reduce carbon emissions and resources through reusing, as a minimum, the shell, and upgrading the plant which in older buildings contributes disproportionately more carbon emissions.

Sustainable refurbishment can avoid the wholesale waste of resources that comes about when swathes of older building stock are cleared to make way for new. Importantly, it also preserves the character of our towns and cities, making them interesting places in which to live.

A more positive perception of refurbishment must be generated in which its wider role in financial and social sustainability is better understood, alongside its environmental benefits. Innovative ways of using sustainable refurbishment to upgrade buildings and bring stock back onto the market must be developed. New ways must be found to harness the close interest that private owners take in their properties and to help them to recognise and adopt sustainable refurbishment as a practical and cost-effective option.

Although there are various predictions over the renewal rate of the existing building stock, most standing today will still exist in 2050. Therefore, it is essential that this stock is viewed positively as a platform for achieving sustainable benefits rather than as a national problem. Millions of buildings have options for upgrading – the trick is to use best practice to identify the right solutions and deliver them successfully for owners and communities.

The various sustainability impacts affecting building refurbishment are described in Table 2.1. This is not an exhaustive list but is meant to show the types of activities and the impacts that will affect each category. The table is

Table 2.1: List of sustainability issues affecting refurbishment projects

Sustainability Category	Typical Measures Affecting Refurbishments
Management	• Formal certification to recognised schemes, e.g. LEED or BREEAM; • Commissioning of the facility; • Use management systems that enable the business unit to manage and improve its social and environmental impacts; • Develop a project brief capturing indicators, targets, policies and management processes that integrate sustainable development; monitor and evaluate their success; • Review the role, and challenge the record, of the principal contractor and key parties of the project team; • Publish information about the environmental and social impacts of the refurbishment two years post-completion.
Emissions to Air	• Minimise sources of air-pollution emissions in the design; • Separate sources of pollution from sensitive receptors; • Use low-emission finishes, construction materials, carpets and furnishings; • Reduce emissions to air, water and land of greenhouse gases, and ozone-depleting substances.
Land Contamination	• Land contamination either recent or historical; • Maximise fiscal incentives to rectify; • Management of any contaminated land; • Determine extent of necessary rectification.
Workforce Occupants	• Reduce nuisance, e.g. noise, odour, visual impact; • Promote flexibility of space and better utilisation of the building; • Encourage people to be involved in giving their ideas and making decisions; • Utilise pre- and post-occupancy evaluation to identify and incorporate facility benefits; • Management of dust and noise during demolition and construction phases of the project.

Sustainability Category	Typical Measures Affecting Refurbishments

Local Environment & Community	• Contribute to the local environment and to the local infrastructure to improve safety and provide opportunities for people to become healthier;
	• Work in partnership with local communities:
	  ◦ Promote and support community activity and volunteering;
	  ◦ Share resources, e.g. equipment, knowledge, skills, and buildings, with community, voluntary, educational or charitable groups;
	  ◦ Take part in schemes that utilise the potential contribution to society of people who might otherwise not be employed, e.g. the long-term unemployed, disabled people, and that help all individuals to contribute to society and realise their potential;
	  ◦ Contribute to the prosperity of the wider local, regional or national economy.

Life cycle of building/ products	• Inclusion of environmental issues throughout the life cycle of a project to include ongoing issues once designed; • Use of materials aligned with the life of the facility; • Design for easy dismantling and demolition.
Energy Management	• Upgrade of plant and BMS to reduce energy consumption; • Generate or use renewable energy; • Increase air tightness in the building to reduce heat loss and draughts; • Revise lighting types and strategy to better utilise daylight; • Review of service-charge payment to influence reduction in utility consumption.
Emissions to Water	• Review potential for sustainable drainage system to meet forward rainfall patterns, e.g. porous pavements, swales and basins or green roofs; • Upgrade key areas of water use, including toilets, kitchen facilities to reduce consumption; • Provide flood defences and methods for their protection.
Use of Resources	• Use low-embodied carbon materials as a preference, which becomes more significant as building use decreases; • Increase use of renewable and recycled materials; • Storage of chemicals and fuel oils.
Waste Management	• Reduce waste of all kinds; • Legal compliance with the WEEE Directive, waste disposal; • Legal compliance with solid and liquid wastes; • Disposal of hazardous wastes such as asbestos; • Management of wastes produced during the refurbishment.
Marketplace	• Use purchasing power to support the local economy and to support organisations that are contributing to sustainable development; • Increase the flow of information to stakeholders; • Increase consultation with stakeholders and use it to inform decision making; • Conduct business ethically, and actively oppose corruption and unfair practices; • Payment of contractors on time.
Human rights	• Promote good health and safety at work, in (global) supply chains and more widely; • Promote skills and activities that help people live more healthily.
Biodiversity	• Protect and enhance native species and their habitats; • Increase the proportion of resources that are from sustainably managed sources, e.g. not over-fished, over-harvested, clear-felled or taken from the wild.
Transport	• Use of video and teleconferencing for meetings; • Management of deliveries to reduce the number of journeys and vehicles used; • Distance travelled by suppliers and contractors to provide services.

based upon a set of 14 themes of sustainability, which are recurring trends from recognised management tools, including the Global Reporting Initiative, social and environmental management standards and Business in the Community[28].

The table can be used as an initial checklist to ensure that all the relevant areas have been assessed and that those that are viable have been taken forward. Some of the measures described below will be applicable across all the levels of refurbishment, such as those affecting workforce occupants, as every refurbishment project will cause an element of change and therefore potential distress to those working in the vicinity. Conversely, some areas involving principal contractors will be applicable only to larger-scale refurbishments.

An approach that has been developed to address these various themes in part is the Prupim framework, which covers the environmental criteria predominantly where greater control is available and demonstrable results are easier to obtain. What the framework does provide is a set of guiding principles or minimum criteria that each project must achieve. Interestingly, few absolute targets are set – only energy and materials – with most of the criteria focusing on how decisions will be made.

Sustainable Refurbishment: A Framework for Decision Making[29]

All refurbishment projects will present a unique combination of aspirations and constraints that will result in unique design and construction responses. There are a number of guiding principles that are to be applied in all cases:

- Avoid over-specification; seek to achieve the optimum design response to anticipated end-user requirements;
- Agree targets and objectives and continuously refine these as the project evolves;
- BREEAM assessments will be undertaken on all projects with a construction spend above a fixed threshold, with the intention that a rating of 'Very Good' be obtained;
- For commercial offices, energy-consumption performance standards shall be targeted to be 25 per cent better than required by Part L2 of the Building Regulations 2006. For other projects, appropriate targets shall be set that shall be not less than 15 per cent better than required by Part L2;
- Adopt passive measures in preference to active solutions and 'low tech' proven technologies rather than innovative unproven solutions;
- Utilise a value hierarchy and focus on measures that deliver the greatest benefit for each pound invested;
- Regulatory-compliant, best-practice solutions are the minimum acceptable standards, and the aspiration in all cases is to exceed these standards;
- In the development of the Waste Management Action Plan, target 80 per cent (weight/volume) of the non-hazardous construction waste and demolition waste for either recycling or reuse;
- For materials and products, establish a target of 15 per cent (weight/volume) whereby their contents are manufactured from either a recycled, or a reclaimed, source.

Endnotes

1 *Silent Spring*, Rachel Carson published by Boston Houghton Mifflin Company (2002)

2 www.ecolo.org/lovelock/what is_Gaia.html

3 Our Common Future, Report of the World Commission on Environment and Development 1987 (A/42/427)

4 www.unglobalcompact.org

5 'Notes on the World POPClock and World Vital Events'. US Census Bureau. http://www.census.gov/ipc/www/popwnote.html

6 'World Population Prospects:The 2008 Revision'. Population Division of the Department of Economic and Social Affairs of the United Nations Secretariat. June 2009. http://www.un.org/esa/population/publications/popnews/Newsltr_87.pdf

7 http://esa.un.org/unpp/

8 *State of the World 2004,* The Worldwatch Institute

9 http://www.ipcc.ch/publications_and_data/publications_ipcc_fourth_assessment_report_synthesis_report.htm

10 *HM Treasury, Long-term global economic challenges and opportunities for Europe,* March 2005, www.hm-treasury.gov.uk/documents/international_issues/int_global_index.cfm

11 *HM Treasury, Long-term global economic challenges and opportunities for Europe,* March 2005, www.hm-treasury.gov.uk/documents/international_issues/int_global_index.cfm

12 http://www.abi.org.uk/Media/Releases/2010/11/Massive_rise_in_Britains_flood_damage_bill_highlights_the_need_for_more_help_for_flood_vulnerable_communities_says_the_ABI.aspx

13 Stern Review: Economics of Climate Change, October 2006, http://www.hm-treasury.gov.uk/sternreview_summary.htm

14 *Climate Change – The UK Programme*, Department for the Environment, Food and Rural Affairs, UK Government 2006

15 http://www.ipcc.ch/publications_and_data/publications_ipcc_fourth_assessment_report_synthesis_report.htm

16 WWF Living Planet Report 2010, www.wwf.org.uk/filelibrary/pdf/livingplanet2010.pdf

17 Ecological Footprint Atlas 2010, Global Footprint Network, October 2010

18 UN Food and Agricultural Organisation

19 Waste Indicators – http://themes.eea.eu.int/Environmental_issues/waste/indicators/ THE EUROPEAN ENVIRONMENT STATE AND OUTLOOK 2010 MATERIAL RESOURCES AND WASTE

20 http://www.grid.unep.ch/waste/html_file/22-23_municipal_rise.html

21 *Policies for Sustainable Consumption; A report for the Sustainable Development Commission*, Tim Jackson, Centre for Environmental Strategy, University of Surrey and Laurie Michaelis, Environmental Change Institute, Oxford University

22 UN Food and Agriculture Association, http://www.fao.org/nr/water/aquastat/data/query/results.html

23 UK Environment Agency, www.environment-agency.gov.uk

24 Rainforest Action Network – www.ran.org.

25 Greenpeace – www.greenpeace.org

26 World Resources Institute – www.globalforestwatch.org

27 Greenpeace – www.greenpeace.org

28 Shah's, *Sustainable Practice for the Facilities Manager*, published by Wiley 2007

29 Sustainable Refurbishment: A Framework for Decision Making, Prupim Developments 2009

3 Occupant Evaluation

Much of the focus on developing and delivering a 'green building' is based upon the physical attributes of the facility and the direct asset value. A facility should also provide an environment to encourage and satisfy employees, visitors and customers within the facility. The combination of both factors is difficult, as they can be in conflict with each other through issues such as ventilation rates.

Responses to the climate-change and asset-performance challenges have often focused on technical solutions aimed either at reducing demand – by improving the building fabric, air-conditioning plant or controls – or at extracting energy more efficiently from the remaining fossil-fuel sources and developing renewable energy. However, technical problems are rarely the only, or even the main, barrier to a sustainable refurbishment. Social and cultural issues are often just as important and the roll-out of technical solutions is often impeded by cultural, educational and political obstacles.

There has been a significant level of research into the provision of adequate space, services and equipment to improve satisfaction and thereby increase performance. Whilst the improvement of the environment does have an impact on productivity, it is difficult to ascertain the true value outside of other changes and influences. However, the evaluation of end-user satisfaction can identify ways to improve the working environment at no cost and to minimise reactive maintenance requests.

Sustainable Refurbishment, First Edition. Sunil Shah.
© 2012 Sunil Shah. Published 2012 by Blackwell Publishing Ltd.

Chapter Learning Guide

This chapter will look at the impacts on occupant satisfaction and the mechanisms to identify and provide effective space for improved productivity. This will include identifying the points for changing space requirements and occupancy evaluation:

- Introduction to occupant satisfaction;
- Triggers for a change in the workplace;
- Pre- and post-occupancy evaluations;
- Behavioural change.

Key messages include:

- Healthy buildings can significantly improve workforce performance;
- Employee feedback will help to target key gaps in the refurbishment;
- The refurbishment of buildings can provide a significant opportunity to improve the performance of their occupants.

3.1   Introduction

One of the drivers for refurbishing buildings is to upgrade tired facilities with a view to reducing obsolescence, the amount of time the facility is not let, and to improving employee satisfaction and performance. This is particularly the case for levels 1 and 2, whereby refurbishing the building might take place more frequently on specific floors or areas.

Such a refurbishment is able to provide increased workspace allowing for a higher working density that provides direct financial savings. In addition, a number of non-financial benefits can be delivered, including improved recruitment and retention, reduced sickness levels and also improved productivity.

A workplace review is not simply a case of reorganising space or ensuring regulatory compliance. It represents an important strategic task that can fundamentally change business performance by maximising efficiencies, enhancing productivity, promoting positive behaviour and successfully communicating the company's values and brand. Almost three-quarters of property professionals feel that the workplace has a role to play in helping companies achieve their business plans, but only 63 per cent are convinced that the board recognises the importance of the workplace, and less than half are satisfied that enough is being done to make the workplace contribute positively to the achievement of business objectives[1].

Poorly designed offices are having a major impact on productivity by as much as 19 per cent. The design is also linked to job satisfaction, recruitment and retention with four out of five professionals considering the quality of their working environment very important to job satisfaction and more than one-third stating that working environment has been a factor in accepting or rejecting a job offer. Over half of professionals believe their office has not been designed to support their company's business objectives or their own job function. Personal space (39 per cent), climate control (24 per cent) and daylight (21 per cent) are the most important factors in a good working environment according to those surveyed [2].

Good office design can increase productivity by nearly 20 per cent – and is a crucial factor in job satisfaction, staff recruitment and retention. The cost of providing accommodation for office workers is dwarfed by the cost of their salaries – the effect of the office environment on staff in terms of increased productivity and effectiveness will have a much greater financial impact than the factors influencing the cost of office accommodation. However, it is difficult to define these costs, which will have a significant bearing on the viability of smaller-scale levels 1 and 2 refurbishments.

A report by PricewaterhouseCoopers for the UK Education Department found a 5 per cent improvement from high-quality school buildings where sustainable improvements had been incorporated. Areas highlighted included open areas for play; light colours to create a sense of warmth; community involvement; daylight and natural ventilation; and a move away from narrow, dark corridors[3].

In the United States, the 'West Bend' study by Walter Kroner documented productivity gains from daylighting, access to windows, and a view of a pleasant outdoor landscape at the West Bend (Wis.) Mutual Insurance Company. The performance of clerical workers in a new building, opened in 1991, was compared to that of workers in an old building, to see which group could produce more reports in an allotted time. Employees in the new building were also supplied with individual controls that allowed them to adjust temperature and other conditions in their work environments. According to the study, productivity gains in the new building increased by 16 per cent, with the personal controls alone accounting for a 3 per cent gain. The building also reduced energy consumption by 40 per cent[4].

Another report is the Heschong Mahone Group study, 'Daylighting in schools', which was conducted on behalf of the California Board for Energy Efficiency. The researchers analysed test scores for 21,000 students in 2,000 classrooms in Seattle, Wa.; in Orange County, Calif.; and in Fort Collins, Colo. In Orange County, students with the most daylighting in their classrooms progressed 20 per cent faster in maths tests and 26 per cent faster in reading tests in one year than those with the least daylighting. For Seattle and Fort Collins, daylighting was found to improve test scores by 7–18 per cent[5].

A study of windows and views in seven buildings in the Pacific Northwest found that employees in work areas with windows were 25 per cent to 30 per cent more satisfied with lighting and with the indoor environment overall, compared to those with reduced access to windows[6]. Window views may be especially effective in providing short rest breaks of a few minutes or less, which have positive impacts on performance and attention[7].

Over the years, many mathematical models reflecting the relationship between the thermal environment and workplace productivity have been developed, allowing a forecast of change in productivity in response to thermal environment variations. It is therefore important to ensure that measures are taken to provide a comfortable internal environment for occupants. For maximum effect, incorporate both tangible and non-tangible/more subjective factors that influence productivity, together with the provision of the flexibility for occupants to control their thermal and visual environment.

The existing research on personal control over environmental conditions, especially temperature and ventilation, shows a strong link to enhanced work productivity as well as to comfort and acceptability (Brager and deDear, 1998)[8]. According to Leaman (2000), two things are invariably important in determining workplace productivity[9] and they can be provided by user controls:

- Excellent discomfort alleviation in a basically comfortable environment, especially in summer;
- Meeting perceived needs quickly and with as little fuss as possible.

3.2   Changes in Work Patterns

The changes in work styles and patterns are leading to a move away from the traditional one-person-to-one-workstation arrangement of desks. Some staff are working part time, with shared roles, flexible hours, flexible locations and they might be home-based. The workstation share ratio provides an indication of desk sharing, or alternative ways of working being employed, at a particular location. A typical desk-share ratio for flexible workspaces is 1.3:1 (i.e. 1.3 occupants for each workstation) with the aim of providing:

- 1:1 for fixed static staff;
- 3:2 for flexible static staff;
- 3:1 for mobile staff;
- $7m^2$ net useable space per office.

The provision of the working space should encourage the collaboration of staff, time spent away from the desk and computer, and greater interaction with other staff. The various types of space should support this flexible approach. This will include an open-plan space, encouraging managers to be located together with staff, and a suite of meeting spaces both formal and, importantly, informal;

- Open plan – The team workspace is the fundamental work unit for the majority of staff. This space provides individually owned space with additional shared facilities to support daily work activities;
- Personal office – The creation of individual offices should be discouraged where there are quiet rooms and small meeting rooms that could be used. Where necessary, an office should be provided only for staff with a managerial role whose job requires working on sensitive matters that require a level of confidentiality. Offices should be located towards the central core of the room so as not to restrict the natural daylight and be a shared space when not in use;
- Meeting room – A range of rooms that are bookable and available to all staff. Larger meeting rooms can also have the facility to be used as training rooms if sufficient data and network points and equipment are available for the users;
- 'Breakout' space – These spaces are provided for informal meetings and 'breakout' intended to promote networking and encourage interdepartmental interaction. They also provide core services such as reprographics, library facilities and stationery around which social interaction is encouraged.

A nominated person should be made responsible for these hubs to ensure they are maintained;

- Quiet room – Non-bookable space for individuals to work in privacy, enabling focus and concentration. Layout should allow for up to two quiet users;
- Restaurant – This is the main communal facility providing a series of spaces for 'breakout' and meals/snacks. This includes the provision of soft furniture promoting informal 'breakout'/meeting space to take the pressure off formalised meeting spaces. Loose furniture allows the flexibility to create large meeting spaces for the whole organisation (and the provision of adequate connectivity allows audio-visual aids). Furniture should be provided that allows small-group meetings and large-group gatherings at meal times.

Triggers for a Change in Workplace

Costs associated with changing the workplace and the disruption caused mean that this option should only be taken when necessary. There are a number of trigger points to identify the appropriate time to provide the change, to review their space utilisation and move towards the best-practice model.

Organisational changes

- Organisation redesign based upon changed focus or realignment of business activities;
- Recruitment/retrenchment of staff;
- Management change within location;
- Mergers and acquisitions – to promote integration and culture change;
- More flexible global workforce – for increasing mobility.

Cost-management initiatives and activities aimed at reducing cost:

- Minimising churn activity through the provision of efficient space in the first instance and reducing cost of change;
- Rationalisation of portfolio due to lease expiry, break clauses or mechanisms to alter space usage.

Changes in available technology, enabling a reduction in the number of desks, the occupant footprint or the desk footprint (these may include flat-screen technology, use of laptops etc.)

Occupational-health changes to the use of the workplace to provide an effective space for personnel to work

3.3   Pre- and Post-Occupancy Evaluations (POE)

An occupancy evaluation involves the systematic evaluation of opinion about buildings in use, from the perspective of the people who use them. It assesses how well buildings match users' needs, and identifies ways to improve building design, performance and fitness for purpose.

The Royal Institute of British Architects (RIBA) Research Steering Group (1991) defined post-occupancy evaluation (POE) as 'a systematic study of buildings in use to provide architects with information about the performance of their designs and building owners and users with guidelines to achieve the best out of what they already have'.

The overarching benefit from conducting a POE is the provision of valuable information to support the goal of continuous improvement. Traditionally, many decisions that are made in the design stage of building projects are based on assumptions of how the occupants function and how they use their spaces.

It is essential to engage with, and gather information from, the occupier to improve user satisfaction and the sustainability of the property. Buildings can add substantial value for their owners and their occupiers. Nevertheless, opportunities are regularly missed and value needlessly subtracted because information is not gathered from building occupiers.

The information gathered from a pre-refurbishment study to learn about the way occupiers use their building space can then be fed back into the design, allowing it to be tailored to specific requirements. However, a fine balance needs to be sought between tailoring a building to a single occupier's needs and not reducing the building's market attractiveness.

If the occupier is unknown, information can be gathered from other occupiers in comparable buildings in terms of size and locality. It may be that engagement exercises where the occupier is known might provide sufficient information to help to identify some general themes about tenant desires.

Occupier Engagement

The following tasks should be carried out and information obtained from the comparable building[10]:

When undertaking a refurbishment project, to make the most of the potential to add value, perform the following engagement exercises during the briefing and early design stages:

* Workshops;
* Focus group;
* Interviews;
* Surveys.

When engaging with tenants, customers or employees, identify the following:

* Refurbishment requirements;
* The type of business of the occupier;
* Quantity of heat-generating equipment;
* Usability of the workspace and the preferred layout (open plan or many partitions);
* The level of the user's current satisfaction;
* Major strengths and weaknesses, i.e. the aspects they would like to take forward and the aspects they would like to change, and their expectations of the new premises;
* User expertise;
* The typical working hours to determine when the plant should be started and turned off;
* Holiday periods when there will be no requirement for plant operation;
* For unknown users, no outcome considerations can be made.

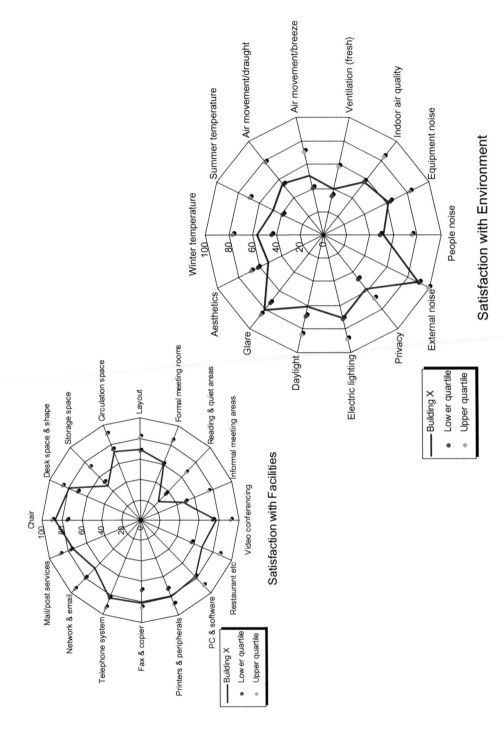

Figure 3.1: Typical outputs from a pre and post occupancy evaluation.

The POE is performed using a questionnaire to gain direct feedback from the occupants, and uses these experiences as the basis for evaluating how a building works for its intended use. It can be used for many purposes, including fine-tuning new buildings, developing new facilities and managing 'problem' buildings. Organisations also find it valuable when preparing for refurbishment, or when selecting accommodation for purchase or rent. Most importantly, the tool allows occupants to provide direct feedback on the performance of the building and how it meets their needs.

A typical output of the POE is described in Figure 3.1 for the two assessments that are performed – one on the facilities provided, and the other on the environment. The assessment for facilities covers questions related to satisfaction with items such as chairs, storage, desks, restaurant and printers. The assessment for environment covers issues including temperature, ventilation, glare, daylighting and noise. Occupants are asked to rank each of these various aspects from 'excellent' through to 'very bad', which enables a consolidated score to be provided. Using similar locations, a benchmark can be provided to compare performance with peer companies.

From this initial point where the current satisfaction is known, the project can incorporate the lessons in terms of which areas to change, and which to improve. As can be seen from the figure, the critical areas to change would include storage space and reading/quiet rooms from a facilities' perspective, and air movement and lighting from the environmental side. The inclusion of feedback from occupants and taking into account their requirements also improve performance simply because issues raised are resolved.

Incorporated within the POE could be a review covering focus groups – small teams who represent the occupants where a greater level of dialogue provides further feedback on the performance and what may or may not work well. A typical POE has three phases:

1. **Preparation:** Identification of user groups, timetabling, selection of participants, letters of invitation.
2. **Interviews:** Small groups of like users are interviewed while walking through the building, which provides the prompt for their comments and observations. A review session is held to verify comments, establish priorities and review the process. Observation studies and written questionnaires may also be used.
3. **Analysis and Reporting:** Documentation of participant findings, generation of recommendations, compilation of a report and presentation. Comparison with the pre-evaluation to demonstrate improvements.

Improving Productivity Through Refurbishments

The BCO-CABE report, 'The Impact of Office Design on Business Performance'[11], makes clear the link between well designed office buildings, business performance and, ultimately, business profitability, and results. Evidence of increases in productivity of 25 per cent from staff who worked

(Continued)

in well maintained, well designed offices were achieved as a result of their relocationdue to building refurbishment; it documents the effects the changes have had on staff, working practices and the bottom line.

Provided below is a case study of a refurbished building where the tenants were in situ, and a research paper on the changes affecting a financial institution.

500 Collins Street, Melbourne, Australia – Kandor Group

The 28-storey, early-1970s building was purchased in 2002 by the Kandor Group with a view to implementing a sustainable refurbishment on a floor-by-floor basis with an in-situ occupancy of 80 per cent being maintained throughout.

A pre- and post-occupancy evaluation were performed on two tenants who moved from original space to newly refurbished space, with a resulting significant increase in productivity of 12 per cent. A number of additional benefits were realised, including a 44-per-cent reduction in the average monthly cost of sick leave in one organisation, and an increase in billings' ratio, despite a 12-per-cent decline in average monthly hours worked from the other organisation.

Staff well-being results related to improvements from indoor environmental quality – improvements resulting from a move to a refurbished floor

Symptoms affecting productivity	Tenant one	Tenant two
Headache	7% reduction	20% reduction
Colds and flu	21% reduction	24% reduction
Fatigue	26% reduction	16% reduction
Poor concentration	20% reduction	5% reduction

In addition, wider sustainability benefits have been achieved, including a reduction of over 50 per cent in energy consumption. The culture of the tenants has also enabled recycling rates to increase significantly from 13 per cent to 42 per cent in a few years, with almost 50 tonnes of dry recyclables being diverted from landfill.

For more details, see: www.resourcesmart.vic.gov.au/for_businesses_ 3661.html

Post-Occupancy Evaluation and Workplace Productivity

Rob Kooymans and Paul Haylock

Centre for Regulation and Market Analysis (CRMA), School of Commerce, University of South Australia

The research concerns four financial-institution branches refurbished between July 2003 and June 2004. Staff and managers completed a questionnaire to determine differences in productivity and attitude between their being in new and old accommodation.

The questionnaire typically used a 7-point Likert scale, enabling the respondent to indicate their degree of agreement with a position or view, including a mid 'neutral' point to indicate a neutral view (1=low, through to 7=high). In addition, a list of 10 features of the branch was provided and respondents were asked to rate the features on a scale of 1–5, indicating the 5 best features of the branch, both before and after the refurbishment: 1 was a measure of the best feature. Respondents were also asked to score the two features that were the least satisfactory in the accommodation, both before and after the works were performed. These scores were recorded by the number of times the item was scored and the average ranking that it was given.

The overall staff satisfaction, before and after the refurbishment projects, shows a significant difference, but what does this mean in practice?

Figure 3.2 shows the 'before' and 'after' scenarios for a range of measures, where productivity and satisfaction with the working environment increased after the refurbishment. However, it was also noted that job satisfaction had decreased, demonstrating the need to separate the physical changes from those that are personnel-based.

The responses revealed a high score on staff views of the image of the branch before the refurbishments, with an initial score of 4.94; however, this grew to 6.31 on the 7-point scale after the refurbishing of the branch, which is a very good outcome from the viewpoint of staff satisfaction with their employer and workplace.

Figure 3.2: Mean productivity, job-satisfaction, and environment-satisfaction scores.

Post-Handover Post-Occupancy Evaluation (POE)

Clients and users are increasingly expecting better performance from the buildings they procure and occupy. The post-handover period is the most neglected stage of any refurbishment project, but it is precisely when much can be fed forward into the completed project for the benefit of the client and the occupants.

Furthermore, much can be learnt, recorded and fed back for reuse on future projects, for the benefit of all the stakeholders.

Of great importance is the closure of the gap between design intent and operation, which commonly exists due to poor commissioning and handover. The use of post-occupancy evaluation can help to identify where the gaps exist, in terms of the operations of the facility, together with the user expectations, to ensure they are effectively aligned. The POE should ideally be performed 12–18 months after the last occupant moves into the facility and should identify the major strengths and weaknesses of a building's performance, providing data that supports the need for, or against, further in-depth evaluation. The aim is to compare the 'big picture' building performance against existing criteria, design intent and the programme.

A POE will provide real information on how buildings perform in use to enable fine tuning so they will perform better, thereby improving the end product. It is the fastest and surest way to improve the economic and environmental performance of buildings and to achieve greater user satisfaction.

An effective POE will result in the collection of evidence from the actual occupants of the building in question. Measured information can be collected from BMS trend logging, energy consumption, temperatures, lighting levels, acoustic performance, maintenance, air movement etc. In addition, survey data is available from design intentions, occupant comfort and satisfaction, facilities-management regime, user understanding of controls, awareness of Building User Guide usage, layout, aesthetics, odours etc.

The POE could be completed by a facilities manager who has knowledge of the building and its HVAC system, and the facilities manager could possibly carry out a POE for the building in question as a routine and regular maintenance item. A 'snapshot' POE every three or five years can quickly recoup costs through problem identification; it can save up to 25 per cent in energy costs with no-cost remedial measures. Alternatively, the building owner could commission a private independent company to perform the evaluation.

For the POE to be successful, data should be used to feed back into design guidelines, criteria and policies and should be carried forward to future projects. Corrective actions for minor problems early in the building's life cycle should be identified, together with those that require further study, and unsuccessful components that should not be applied to future projects.

Where the POE identifies that certain aspects of the building require a more detailed review, an investigation should be performed. These aspects will require more detailed and formal data-collection techniques. Standard questionnaires can be used to survey respondents or, alternatively, customised questionnaires can be prepared. Structured interviews and recording of responses can also be included for analysis, which can supplement the responses to questionnaires. Where significant issues arise, or where a number of issues arise from a single root cause, a high-level investigation should be performed focused on this single aspect.

The data can be used to understand the cause and effect of issues in building performance, to design corrective action plans and to provide information from lessons learnt that can be applied to future projects. For significant issues, it may lead to changing the design criteria.

Endnotes

1 Performance Measurement in Real Estate Usage, part of the Directors Briefing set, Haywards 2005, www.haywardsltd.co.uk

2 Gensler's These Four Walls: The Real British Office 2005, www.gensler.com/

3 *Building Performance: An Empirical Assessment of the Relationship Between Schools Capital Investment and Pupil Performance*; PricewaterhouseCoopers; ISBN 1 84185 402 6; http://findoutmore.dfes.gov.uk/2006/08/school_building.html

4 Kroner, W.A., Stark-Martin, J.A. and Willemain, T., 1992, *Using Advanced Office Technology to Increase Productivity. Rensselaer Polytechnic University*, Center for Architectural Research

5 Heschong Mahone Group study 'Daylighting in schools', www.h-m-g.com/projects/daylighting/projects-PIER.htm

6 Heerwagen, J., Loveland, J. and Diamond, R., 1991, *Post Occupancy Evaluation of Energy Edge Buildings*, Center for Planning and Design, University of Washington

7 Zijlstra, F. R. H., Roe, R. A., Leonora, A. B. and Krediet, I., 1999, *Temporal Factors in Mental Work: Effects of Interrupted Activities*, Journal of Occupational and Organisational Psychology, 72: 163–185

8 Brager, G.S. and deDear, R.J., 1998, Thermal adaptation in the built environment: A literature review. Energy and Buildings 27: 83–96

9 Leaman, A., 2000, The Productive Workplace: Themes and variations, Building Services Journal

10 Way, M. and Bordass, B., 2005, Building Research and Information 33 (4), 353–360. Making feedback and post-occupancy evaluation routine 2: Soft Landings involving design and building teams in improving performance

11 www.bco.org.uk/news/detail.cfm?rid=56

Part 2
Managing Refurbishment as a Process

Sustainable building refurbishment is driven by the market and legislation. With a focus on energy security, costs and climatic change, the impacts of energy have partly been internalised within the cost of both these drivers, enabling a differentiation to be seen in the market. Whether or not the other environmental practices become internalised will have a bearing on driving these areas forward.

How these drivers are packaged together to develop the most appropriate strategy to meet a client's objectives with their buildings, and how they are deployed, are critical to the success of the refurbishment.

This part will cover the overarching management aspects that help to deliver a successful refurbishment project.

Chapter 4 will look at the drivers, both financial and non-financial, that are promoting sustainable refurbishment, together with the underlying legislative requirements to be met and the increasing use of fiscal instruments to point building owners and occupiers in the right direction;

Chapter 5 provides an overview of the key risks and benefits of a sustainable refurbishment project affecting the development of an effective business plan and strategy;

Chapter 6 describes a framework to enable the refurbishment to be delivered through techniques that can be used, including Green Leases and ISO14001. The application of the framework provides a rigid structure to review, implement and, importantly, ensure that the sustainability aspects have been achieved post-completion;

Chapter 7 describes how the sustainability issues can be managed within the project through recognised systems, including LEED and BREEAM. Implications for individuals through occupancy evaluation and behavioural-change requirements are also covered as key issues that predicate the success of the sustainable refurbishment into operation.

Sustainable Refurbishment, First Edition. Sunil Shah.
© 2012 Sunil Shah. Published 2012 by Blackwell Publishing Ltd.

4 Drivers for Sustainable Refurbishment?

The delivery of sustainable refurbishment is ultimately down to the market on a local level and what investors, clients and occupiers are willing to pay. Such an approach is reliant upon the effective valuation of buildings that have incorporated sustainable measures in the capital cost to deliver the refurbishment, together with the financial and non-financial benefits that result.

It will also rely upon a strong forward-policy approach from the government and upon legislation to provide a steer for the future market and likely implications. There has been significant movement in this area globally as countries look to correct for the market failure of energy in particular and of sustainable buildings in general.

Evidence is developing from the market of a shift towards sustainable refurbishment for certain aspects. Globally, 2 per cent of buildings are built as energy-efficient, 'green' buildings and this is set to continue to increase. However, new build only represents less than 2 per cent of the building stock each year, so to put the green buildings' programme in context, it represents a 0.04 per cent change each year. The challenge is how to utilise the market mechanisms and legislation to mainstream the positive start that has been made.

Sustainable Refurbishment, First Edition. Sunil Shah.
© 2012 Sunil Shah. Published 2012 by Blackwell Publishing Ltd.

Chapter Learning Guide

This chapter will provide an overview of the drivers affecting the sustainable-refurbishment market and how the market is responding.

- Market pressures from tenants and customer demands;
- Financial evidence of improved return on investment capturing financial and non-financial benefits;
- Legislation being implemented to measure the performance of buildings and provide effective comparisons;
- Climate change at a global level including Kyoto and the various national agreements on carbon reduction;
- Provision of skilled individuals and organisations to deliver cost-effective, sustainable buildings;

Key messages include:

- There is emerging market evidence that energy-efficient buildings command a higher resale value, and are vacant for shorter periods of time;
- The US has a greater level of credible evidence focused on energy efficiency;
- Legislation on energy has underpinned the internalisation of carbon as a market requirement for refurbishment.

4.1   Market Pressures

There are a number of factors that are affecting the current property market and having an impact on the decisions that those involved in the industry – investors, agents, occupiers – are making. The following section provides an overview of some of the key changes affecting the delivery of sustainable refurbishment. Regional differences will apply, for example, energy-price changes are having a lesser impact in the Middle East than in Europe.

Better Utilisation of Buildings

Commercial buildings constitute a significant contribution of carbon emissions, but are only used to a fraction of their full potential, typically for 30 per cent to 40 per cent of the year. More flexible use of the space, to meet the needs of an evolving workforce, can significantly cut the environmental impact. From a sustainability perspective, the development of mixed-use facilities or those that are open to the wider community help not only to reduce the environmental impact but also to provide a social benefit.

Technology has been completely transformed over the past 50 years but, unfortunately, buildings have not progressed at the same rate. The needs of organisations and people have also changed – the new generation coming into the workplace require more flexibility about how, when and where we work. Consequently, there is a need to rethink the way space is used, away from monolithic, one-organisation, one-use buildings towards more complex, multi-use buildings with a focus on density, intensification and reuse of existing buildings.

This can be achieved by using buildings for different purposes at different times, as with Shell's learning centre in the Netherlands that provides staff training during the week and is used as a hotel at weekends. Similarly, London's Oxo Tower is a mixed-use building which combines commercial, leisure and residential units by housing flats, shops and a restaurant. Renovating existing buildings was key to this, particularly making better use of those in locations that would minimise the environmental impact of their users.

The refurbishment of buildings provides an opportunity to revise the layout and functionality of buildings for a mixed use (including a potential change of use), provides improved flexibility and a move towards 24–7 operation. When buildings are vacant, the opportunity to revitalise the space to make it more attractive to tenants is greater. There is also the recycling of buildings with uses no longer required by the market into buildings with uses that are in demand, for example, residential developments from offices, as has been historically performed. Much property is now held by banks – distressed assets – which can act as a driver in this direction.

Landlord–Tenant Relationship

Many buildings are refurbished speculatively with unknown tenants in mind. Landlords are commonly providing the refurbishment to the lowest specification to attract tenants, enabling them to make any additional changes to tailor their space to meet their requirements. In many markets, air-conditioning is standard and therefore tenants can request that this is included. The implication here is to provide a space that is in keeping with the rest of the market and not different from it.

Such an approach goes against the concept of sustainable refurbishment, whereby a higher rental return would pay for any initial investment. With such a short-term view of assessing the requirements of the next tenant, the minimum is often specified. This is exacerbated by the split incentive in play, whereby tenants pay for energy costs and landlords pay for the building, which reduces even further any incentive for any refurbishment projects to improve energy efficiency or other sustainability measures.

Agents have also been victims of the crossfire in this relationship, being accused of creating a barrier to the uptake of particularly energy-efficient buildings. By focusing on the traditional parameters of location and cost, agents either do not provide, or otherwise mention only briefly, any relevant information about the energy efficiency and other sustainable factors of buildings. As a result, tenants are not able to make informed decisions on potentially 'lower-carbon' buildings. The impact is to reduce the potential uptake of energy-efficient buildings, thus having a negative effect on the sector.

Over the years, leases in most commercial buildings have shifted the payment of operating expenses from the landlord to the tenant. In some cases, this shift is achieved through sub-metering, in which the tenant pays directly for the actual use, such as the payment of electricity bills. In other cases, by the use of triple-net leases, all the building's operating expenses are paid by the landlord and then passed on for reimbursement by the tenant. Even in leases where the landlord pays the base operating expenses, there is usually a provision in the lease that states that increases, or decreases, in operating expenses after the first year are passed on to the tenant. Thus, there is a misalignment of incentives, with the landlord responsible for the capital costs for making the building more energy-efficient, but with the tenants benefiting from the savings.

In a building with a single tenant on a long-term lease, this misalignment can be rectified either by having the tenant pay for the renovation, or by having the tenant agree to pay some, or all, of the savings in additional rent. For this reason, most of the energy-efficiency renovations have been to single-tenant, owner-occupied, or public buildings. In a multi-tenant building, it would be very time-consuming for the owner to negotiate with each of the tenants a lease amendment that enables the landlord to be reimbursed in some way if the building consumes less energy. It would also be hard to explain this surcharge to a potential new tenant.

The growth of 'green leases' (see Section 6.3) has provided a mechanism to address this problem by providing incentives for both parties to manage and reduce resource consumption.

Utility Costs

Crude oil, natural gas and other fossil fuels still contribute the majority of energy that business and consumers use. The increasing cost of these fossil fuels is causing a number of economic implications for the higher-energy users – aircraft prices have increased, together with fuel surcharges that are levied in addition to pay for the increase. This increase has been combated by the growth of low-fuel aircraft, which are more efficient and help to reduce costs.

Similarly, many industrial facilities have also been affected and they are investing in alternative technologies to consume less energy and, in many cases, generate their own. Data centres are utilising natural ventilation and being located in places with a colder climate to make use of this 'free' cooling.

With utility costs in US office buildings averaging about US$2.50 per square foot (US$26.90 per square metre), a 20 per cent reduction in energy use yields savings of 50 cents per square foot (US$5.40 per square metre). Even if the owner could keep 100 per cent of those savings and even if it only cost US$3 per square foot (US$32 per square metre) in renovation costs to achieve those savings, the building owner would only receive an extra US$50,000 in income in a 100,000-square-foot (9,300-square-metre) building. Contrast that US$50,000 with the extra revenue and profit that the landlord could achieve if they spent this same amount fitting out and leasing 10,000 square feet (930 square metres) of office space. The new office lease might result in five times as much income flowing to the bottom line.

However, for many low-energy consumers, including those in office environments, the increase in energy prices is still having a limited impact and not driving behavioural change in the utilisation of the building or in the incorporation of low-energy fixtures during a refurbishment. In many countries in the Middle East, energy is subsidised and is passed on to the commercial organisation at limited cost. A project by MED-ENEC achieved a 57 per cent reduction in energy use for a facility in Algeria, which equated to a saving of €225 per annum. The cost of energy for many organisations represents only about 1 per cent to 6 per cent of business-operating costs, in contrast to staff costs of some 85 per cent. The potential savings are therefore relatively small when many of the same organisations are targeting growth at 8 per cent per annum.

Behaviour is starting to change as a result of the increased energy prices affecting those who pay for their own energy, which can include owner-occupiers, as well as commuting personnel. The 2008 spike in energy prices effectively killed the Hummer-type vehicle and led a move towards energy-efficient and hybrid vehicles. Further spikes are likely to drive similar behaviour from those who are most affected by the changes.

Nevertheless, the cost of energy would need to increase dramatically to have a meaningful impact on the behaviour of many landlords.

Tenant Demands

The demand from tenants for more efficient buildings that can reduce operational costs has always been in play and has recently become linked with a recognition of the benefits from occupant satisfaction.

A recent assessment of the market concluded that developers may look to cut costs by incorporating more modular, 'cookie-cutter', streamlined designs, offering different exterior finishes to tenants. The future promises more value-oriented development, not ostentatious projects. Tenants will emphasise function and efficiency, and green, energy-saving sustainability features will be expected[1]. Energy is now a 'front and centre' issue for tenants choosing office space, and becoming 'unavoidable' for the sector. Green buildings currently comprise just two per cent of the market, but respondents expected them to become increasingly sought after as tenants look to cut operating costs and support corporate environmental and social responsibility goals. As a result, sustainable buildings will be positioned as being healthier, more attractive and more marketable.

The UK Occupier Satisfaction Index 2008, by Kingsley Lipsey Morgan and IPD Occupiers, measures satisfaction among customers of the UK commercial-property industry. When asked what occupiers wanted from the property industry in 2008, the survey respondents cited lease flexibility, relationship building, responsiveness to requests, value for money, transparency and sustainability.

The report stated: 'Green issues are becoming more important to occupiers and are playing a bigger part in building choice. While occupiers welcome the property industry's progress on environmental issues, they feel that there is scope for faster progress. Occupiers recognise that awareness of green issues has significantly increased within the property industry, but they now want to see the talk translated into action. Some believe that occupiers and legislation are driving the change and that the property industry needs to be more proactive.'

A 2008 Jones Lang LaSalle/CoreNet Global survey of over 400 occupiers around the world reported that 69 per cent of corporate-real-estate executives said that sustainability was a critical business issue. Forty per cent rated energy and sustainability as major factors in their companies' location decisions. Forty-two per cent said they were willing to pay a 1 per cent to 5 per cent premium to lease green space and 53 per cent said they would pay a premium to retro-fit property they own to gain sustainability benefits.

It is recognised that although green buildings do not necessarily attract higher rents, sustainable buildings lease well in a weak market, with emerging evidence that tenants are willing to pay more for some green characteristics. In addition to offering lower costs, sustainable office space engendered smoother relationships between real-estate firms and planners, and gave businesses an advantage when recruiting, as candidates are increasingly taking environmental policies into account when making employment decisions. In this way, the investment decision is not based on environmental issues alone, but takes a broader view across the economically driven business benefits.

However, whilst developers, government organisations, shareholders and occupiers are more concerned about sustainability, many fund managers remain sceptical that the higher upfront costs of sustainable property will actually lead to financial rewards.

A report from CoreNet Global and Jones Lang LaSalle[2] shows that business leaders are focusing on balancing the environmental performance of their buildings with overall financial performance. The report showed that 92 per cent considered sustainability criteria when choosing business real estate. Half said that they would pay more for green office space. A quarter said that they would pay more in rent if it were offset by lower energy costs. The report said that this willingness might have resulted from the relatively stable economic climate, compared to the previous two years. The survey also saw a shift from previous years in attitude towards employee health and productivity – 31 per cent ranked it as the most important measure of success.

Recent research[3] showed an emerging and increasing demand for sustainable offices in the UK, but location, availability of stock and other factors continued to remain more important in determining occupiers' final choice of office. The research, which was based on the actual moves of 50 major corporate companies, also found that the most important barriers in the market were seen as being length of payback period; the initial investment costs of new technology; and lack of supply. The research also found that corporate occupiers were often looking for a payback of less than five years for a sustainable, energy-efficient building. This seems to support the view that longer payback periods can create problems in terms of cost-effectiveness as opposed to shorter tenancies.

The challenge is to make the refurbishment of existing stock cost-effective so that, on the one hand, energy-efficient property can attract tenants with a rent that they are prepared to pay, and save energy costs within a viable payback period, but, on the other hand, give the investor an enhanced rent and lower risk for the capital outlay that is being made.

So how can the remaining barriers be overcome? A key problem seems to be that we are still relying on market forces to resolve what is a wider societal problem, and this is, of course, accentuated in the current property-market downturn. Green leases can help overcome some of the landlord–tenant incentive-based issues, but ultimately a key test remains: how to prove the value proposition of sustainable commercial property. For occupiers, this means recognising the benefits of sustainable buildings that can offer lower energy bills, better productivity and improved CSR credentials. If there is evidence of a market for such buildings driven by demand and supply, then valuers will start to reflect sustainability in their valuations, and buildings with established sustainable credentials will start to show a differential in comparison with their 'conventional' peers.

However, this might take many years to filter through the system naturally, as evidence will be required to substantiate the valuations being made.

Underpinning this should also be a focus on 'whole-lifecycle costing', or related techniques, which can help key stakeholders attribute the costs and financial benefits of such buildings, identify how these vary over time, and compare and evaluate new-build and refurbishment options. If a code for sustainable buildings applied to both new and existing commercial property, there would be more tangible progress, particularly if this was coupled with improved incentives for the refurbishment of commercial property.

Investors should also consider the difference between sustainable and energy-efficiency projects, and the different costs and benefits they are likely to bring. Whilst green and sustainable projects might increase the value of the building to certain tenants at the time of the sale or lease, they might not decrease operating costs. Energy-efficiency improvements are more likely to impact both operational savings and resale premiums[4].

4.2   Return on Investment

'95% of unlisted UK and continental European real-estate fund managers believe **there is a relationship between environmental performance and financial returns'**, according to a survey sponsored by Aviva Investors and the Environment Agency Pension Fund in May 2009[5]. 'However, the majority felt this relationship was difficult to quantify at the current time.'

There is emerging evidence in this area through a number of individual case studies from which we can form strong opinions about the benefits of green investment. In addition, there are a few fund managers looking at performance within their own portfolios to determine the effects and influence. However, for a number of reasons, there are very few large-scale studies analysing data from which a robust response can be provided. From a practical perspective, there is limited good-quality data available to analyse, largely on account of the relative newness of the debate. A further issue relates to the lack of a standard definition of a 'sustainable property', leading to difficulties in comparing data.

Most of the data being collected and analysed originates from the US through the Green Star and LEED programmes, which enable the collection of comparable data sets. There has been difficulty in obtaining data from elsewhere, largely from the lack of robust and consistently maintained data sets. It is also noted that the only data that is evaluated relates to energy costs and consumption, rather than wider environmental impacts.

A RICS study, 'Doing well by doing good?'[6], provides detailed research on the US real-estate market. Data was collected through Energy-Star-registered and LEED-registered buildings, which identified 1,360 green office buildings, of which 286 were certified by LEED, 1,045 were certified by Energy Star, and 29 were certified by both LEED and Energy Star. Of this total, 893 buildings had sufficient data to enable effective analysis, comprising 694 with rental data as of September 2007, and 199 that were sold between 2004 and 2007.

The report identified that there is a premium for the rents that green buildings with the Energy Star rating can command, but no premium could be found for LEED-rated buildings. Obviously, it varies according to a number of factors, but the aggregate premium for the whole sample is in the order of 3 per cent per square foot compared with otherwise identical buildings – allowing for the quality and the specific location of office buildings. This implies that the green buildings were let more of the time. When looking at effective rents – rents adjusted for building-occupancy levels – the premium is even higher, above 6 per cent. The researchers also looked at the impact on the selling prices of green buildings, and here the premium is higher again, in the order of 16 per cent. What is implied is that the upgrading of the average non-green building to a green one would increase its capital value by some US$5.5 million, not including the cost of conversion.

An analysis of the relationship between actual energy use in a building and its financial performance reveals outperformance by energy-efficient buildings. An increase of 10 per cent in the energy efficiency of a green building is associated with a 0.2 per cent increase in effective rent. The analysis also shows that a US$1 saving in energy costs from increased thermal efficiency yields a return of roughly US$18 in the increased valuation of an Energy-Star-certified building.

The study provides some evidence of the economic value of certifying sustainable buildings in the commercial sector. The economic argument for having an energy-efficient building is strong. Any business wishing to maximise profits will have to start looking at increasing the energy efficiency of its buildings in order to remain competitive.

The RICS claimed that the results suggested that tenants and investors would pay more for an energy-efficient building, but not for buildings that were sustainable in a broader sense. The growing evidence of economic benefits from energy efficiency means that non-green buildings will eventually become an outdated model. What this means for the current market is that other than energy, no sustainability measures are taken into account in commercial-property valuations.

A study by Arup and Davis Langdon, for the Property Council of Australia, of three different types of building has shown that sustainable refurbishment produces a return of better than 10 per cent on investment[7]. The study analysed three generic building types – a 20,000-square-metre CBD tower, a 15,000-square-metre city-fringe high-rise and a 2,800-square-metre suburban office, all built in the 1980s. The analysis documented the significant return on investment from sustainable refurbishment, and suggested that the impact on the property market of the Federal Government's Carbon Pollution Reduction Scheme (CPRS) would make this return even more attractive.

The results provide clear evidence that upgrading an established, existing building to achieve a minimum 4.5-star National Australian Built Environment Rating System (NABERS) rating would provide a positive return on investment whilst ensuring that the asset remains competitive with contemporary buildings.

From an investor's perspective, studies show that returns on green and energy-efficiency features and initiatives are indeed positive, but do not show significant outperformance on a portfolio basis. Pivo and Fischer show that investors who purchased a portfolio consisting solely of green office properties over a ten-year period between 1998 and 2008 would have earned only a slightly higher return at slightly lower risk compared to a portfolio without green features[8]. This small degree of outperformance might be due to the unclear nature of the 'green premium' and how green features are priced in the market. Academic research has shown that the premium is dependent on the size, location and maturity of the green-building market in specific regions. Further research is certainly needed, but it is likely that climate-change legislation and additional government man-dates and incentives will increase the supply of green buildings and make brokers and other intermediaries more aware of the potential value of such properties[9].

Energy-saving measures are still taking a back seat for property investors, according to a new report published by three European institutional investors. The report by Universities Superannuation Scheme, APG Asset Management and PGGM Investments revealed that despite the fact that energy-saving investments can create value, the majority of the 700 listed property companies and fund man-agers surveyed are not yet actively managing environmental issues in their portfo-lios[10]. The survey also found that a 'strikingly low number' of property companies are able to report actual figures on their energy consumption (19 per cent of respondents), water consumption (16 per cent), or carbon emissions (14 per cent).

As a result of the research, the parties created a new environmental bench-mark, the 'Environmental Real Estate Index', which includes scores on environmental-management practices and on the implementation of these prac-tices. The index allows institutional investors to compare the environmental score of individual property investments with their environmental, real-estate targets. The intention of this benchmarking is to serve as a catalyst for environmental engagement in real-estate investments (Figure 4.1).

Benchmarking the energy consumption of a real-estate portfolio is the key first step to making properties more efficient, and the current lack of metrics indicates that we are standing just at the beginning of the road to energy efficiency in the commercial, real-estate sector. The environmental performance of the property sector is likely to improve, as many of the property companies surveyed indicated that the assets acquired or developed in 2008 comply with green or energy-efficiency standards, such as LEED or BREEAM.

Owners of commercial buildings in the US could save more than $41 billion a year in energy costs, if all currently existing commercial space were placed in a decade-long, energy-efficiency, retro-fit programme requiring an annual invest-ment of about $22.5 billion, according to a new report by Pike Research[11].

The report acknowledges that whilst the figures are impressive, they reflect the **market potential** for energy-efficiency retro-fits – rather than the actual market, which under current conditions is a fraction of the potential. The building retro-fit industry faces a number of key challenges. The current financial crisis has had a significant dampening effect on property owners' investments in their properties. Financing for such projects is scarce, and the limited investment in building efficiency is not keeping pace with the growing national demand for energy.

Private, commercial buildings present the largest untapped opportunity for energy-efficiency retro-fits and account for nearly all existing commercial space. In contrast, federal, non-industrial buildings comprise less than 3 per cent of existing commercial space, but major retro-fits in federal facilities and other institutional buildings are far more likely to receive funding than are projects outside the sector.

If the goal of the energy-retro-fit industry is to spend a little money on effi-ciency, whilst total national demand for energy continues to grow, then present policy is functioning well. However, if the goal is to reduce the total demand for energy in buildings over time, by the 50 per cent or more, needed to address international competitiveness, global warming and energy independence, then present energy policy needs to be substantially reformed.

Display Energy Certificate
How efficiently is this building being used?

HMGovernment

A Government Dept
12th & 13th Floor
Jubilee House
High Street
Anytown
A1 2CD

Certificate Reference Number:
1234-1234-1234-1234

This certificate indicates how much energy is being used to operate this building. The operational rating is based on meter readings of all the energy actually used in the building. It is compared to a benchmark that represents performance indicative of all buildings of this type. There is more advice on how to interpret this information on the Government's website www.communities.gov.uk/epbd.

Energy Performance Operational Rating

This tells you how efficiently energy has been used in the building. The numbers do not represent actual units of energy consumed; they represent comparative energy efficiency. 100 would be typical for this kind of building.

More energy efficient

A 0-25

B 26-50

C 51-75

D 76-100

100 would be typical

E 101-125   ◄108

F 126-150

G Over 150

Less energy efficient

Total CO₂ Emissions

This tells you how much carbon dioxide the building emits. It shows tonnes per year of CO_2.

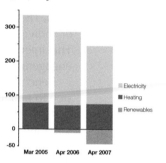

Electricity
Heating
Renewables

Mar 2005   Apr 2006   Apr 2007

Previous Operational Ratings

This tells you how efficiently energy has been used in this building over the last three accounting periods

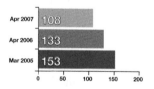

Apr 2007   108
Apr 2006   133
Mar 2005   153

0   50   100   150   200

Technical information

This tells you technical information about how energy is used in this building. Consumption data based on actual readings.

Main heating fuel:	Gas
Building Environment:	Air Conditioned
Total useful floor area (m²):	2927
Asset Rating:	92

	Heating	Electrical
Annual Energy Use (kWh/m²/year)	126	129
Typical Energy Use (kWh/m²/year)	120	95
Energy from renewables	0%	20%

Administrative information

This a Display Energy Certificate as defined in SI2007:991 as amended.

Assessment Software:	OR v1
Property Reference:	891123776612
Assessor Name:	John Smith
Assessor Number:	ABC12345
Accreditation Scheme:	ABC Accreditation Ltd
Employer/Trading Name:	EnergyWatch Ltd
Employer/Trading Address:	Alpha House, New Way, Birmingham, B2 1AA
Issue Date:	12 May 2007
Nominated Date:	01 Apr 2007
Valid Until:	31 Mar 2008
Related Party Disclosure:	EnergyWatch are contracted as energy managers

Recommendations for improving the energy efficiency of the building are contained in Report Reference Number 1234-1234-1234-1234

Figure 4.1: Example of a Display Energy Certificate. Source: Department of Communities and Local Government.

If code policy, design tools, financial incentives, and regulations focus on energy efficiency at the following intervention points (as identified by the non-profit research organisation, Architecture 2030), the incremental cost of efficiency will be very small:

- Building design – schematic design, material and building systems' selection;
- Existing building purchases;
- Leasing/tenant improvements;
- Building-renovation cycles;
- Rebuilding (after a natural disaster).

Programmes that do not recognise these intervention points, or fail to take advantage of them, face unnecessary obstacles, costs, and potential failure. A national carbon-trading system could have a major effect on the retro-fit market. If national carbon-emissions legislation addressed energy use in commercial buildings with a combination of high energy prices and reinvestment incentives, then the market for energy-efficiency retro-fits (and the education of the workers in this market) would explode with activity.

One of the most comprehensive statistical analyses to date identified that US buildings labelled under the LEED or Energy Star programmes charge 3 per cent higher rent, have greater occupancy rates, and sell for 13 per cent more than comparable properties[12]. Labelled buildings have effective rents that are almost 8 per cent higher than those of otherwise identical, nearby, non-rated buildings. However, whilst projects certified under LEED for Existing Buildings was the largest group, it was not distinguished from other registered and certified buildings within the LEED for New Construction and LEED for Core and Shell projects.

The interaction between LEED and Energy Star ratings was assessed, with the benefits of energy efficiency and LEED's broader sustainability metrics both being 'fully capitalized into rents and asset values'. Data on nearly 2,700 buildings that were certified, or were pursuing certification, through LEED or Energy Star, was obtained. Of those, 1,943 were rental properties, and 744 buildings provided sales figures. The study matched each green property with conventional office buildings located within a distance of one-quarter of a mile, using the CoStar database of commercial real-estate information, and corrected for variables such as age, size, quality, number of storeys, date of last renovation, the presence of on-site amenities, and proximity to public transport (see Table 4.1). In total, nearly 27,000 buildings were analysed.

Whilst this evidence is compelling, statistically valid data on the cost of attaining LEED and Energy Star ratings is not yet available, but anecdotal evidence

Table 4.1: Occupancy rate and cost return for commercial real estate (Source CoStar Group)

Building type	Occupancy rate	Rental rate per ft^2	Sale price per ft^2
ENERGY STAR Certified	91.5%	$30.55	$288
Non-ENERGY STARI peers	87.9%	$28.15	$227
LEED certified	92.0%	$42.38	$438
Non-LEEd peers	87.9%	$31.05	$257

suggests that any green cost premium is lower than the increase in value of the properties.

The value of green retro-fits is well established. According to McGraw Hill, 'today green building comprises 5 to 9 percent of retrofit and renovation market activity by value – projected to grow to 20–30 percent in just five years', based on the reduction of new construction and the volume of existing buildings – and that appears to be an average number. In major metropolitan areas, the market is far greater.

In New York City, PlaNYC refers to the fact that '85 percent of the buildings that will exist in 2030 are already here today'. Representatives at PlaNYC further identify the importance of the retro-fit of these facilities to meet energy-reduction targets for the city of New York and the city's multi-step process to ensure that owners meet those targets. Of course, shrewd owners know that there is less risk in green retro-fit as the performance indicators are already established.

The search for a link between the sustainability credentials of a building and its rental, and/or its capital, value began some ten years ago, but is still in its infancy. Early attempts at presenting a business case were founded on low additional cost, lower risk and reputational benefit. There was also a strong argument that cost savings in the hands of the tenant would result in rental differentiation, leading in turn to a reduction of long-term risk and better 'future-proofing' of investments. The argument also turns on the ability to support investors' and occupiers' CSR policies.

Many surveys have been undertaken that give credence to the view that sustainable buildings are worth more. Recently, data has begun to emerge in the form of a handful of large-scale studies based on the US office market. The evidence is acknowledged still to be tenuous and generally goes no further than to point to a connection between higher rents achieved for LEED-accredited and Energy-Star-accredited buildings compared with similar non-accredited buildings. There is no substantive evidence that demonstrates any firm connection with increased capital values achieved on sale.

Currently, more empirical studies are hampered by the lack of agreement as to what constitutes a sustainable building and the lack of a simple benchmark that remains static over time. Moves towards the agreement of a universal definition are afoot and as this starts to filter down to the market, and as the work of organisations such as IPD begins to provide data on sustainable buildings, then a sounder basis for analysis will emerge.

For now, the value and sustainability link is argued strongly in theory and as an opinion, but in terms of hard evidence, it is very limited and restricted to rental differentiation within a tight geographical area and within one sub-sector of the market[13].

4.3   Regulatory Incentives

Environmental impacts have a significant effect on the market in that the impacts and resulting costs are not part of the price of the product being sold. The cost of supplying energy, metals or consumables does not include the additional costs related to the climate-change implications arising from the increase of greenhouse gases. From an economics perspective, carbon is seen as an externality.

Over the past two decades there has been a drive to better understand the costs of climate change, and the way in which these costs can be addressed within the market as a whole. The first global agreement was the Kyoto Protocol, signed in 1997 but only ratified in 2005, followed by the EU Emissions Trading Scheme in 2005. Since then, the EU has been actively developing market-based financial incentives and a regulatory minimum for operations within the EU block. More recently, the US has been promoting measures as part of their financial rebuilding. Standards to reduce energy and promote sustainable practices include Top Runner in Japan, Eco Design directive in the EU and a number of programmes in Australia.

The section below provides an overview of the significant pieces of legislation that will affect the sustainable refurbishment of buildings. The key trends that are being seen globally are:

- Legislation is being widely adopted across many countries and individual states, requiring sustainability targets and thresholds to be met as a minimum requirement;
- Incentives are increasingly being used to meet higher targets, including rebates, property and corporate-tax incentives, density bonuses and expedited permitting;
- The demand for transparency in a building's energy performance is growing, requiring developers to state the performance of their building, enabling tenants to assess its energy performance before making a decision.

EU Revised Energy Performance of Buildings Directive (recast 2010)

The recast is an update of EPBD, 16 December 2002, which set out to promote the efficient use of energy, to meet the Kyoto Protocol and to maintain the global-temperature rise below 2°C. What it has achieved is to provide a regulatory framework across the European Union to implement building energy-efficiency standards and practices for, amongst other areas, medium-level to major-level refurbishments. The directive is enacted within each member state for all new buildings with areas larger than 1000 square metres or for refurbishments affecting more than 25 per cent of the property.

The 2002 EPBD sets a number of requirements[14]:

- Minimum energy-efficiency standards covering major energy-consuming items, including heating, cooling, lighting, ventilation and hot water;
- Regular inspections of air-conditioning systems above 12 kW and of boilers with an output above 20 kW; all inspections must be carried out by independent, qualified experts;
- Energy-performance certificates – see later in this section.

The key changes resulting from the EPBD Recast 2010 are:

- Each country must set a minimum energy requirement for existing buildings and also for when they are renovated, retro-fitted or replaced; a renovation is typically where more than 25 per cent of the building is rebuilt or altered;

- National plans should increase the number of 'nearly zero' energy buildings (which may include generating 'green' energy) and report progress by December 2012;
- Energy-performance certificates or display-energy certificates (either may be used) must be displayed in a prominent place;
- All countries must now regularly inspect air-conditioning systems;
- By January 2013, all states must report on effective penalties for non-compliance.

Energy Labelling and Performance of Buildings Directive (ELPBD)

The EU agreed to alter their building codes so that **all new buildings constructed from the end of 2020** meet higher energy-saving standards and, to a large extent, use renewable energy.

Existing buildings will have to be upgraded, where possible, during major renovations. When renovating, owners will be encouraged to install 'smart-meters' and replace heating, hot-water plumbing and air-conditioning systems with high-efficiency alternatives such as heat pumps. Regular inspection of boilers and air-conditioning systems will be required in all, not just some, EU countries. Public authorities' building projects are to lead the way two years earlier. Part of the funding for these changes will come from the EU budget.

The labelling colour scheme – ranging from dark green for the most energy-efficient products to red for the least energy-efficient ones – will be adjusted accordingly, so that the highest energy-efficiency class will remain dark green and the lowest energy-efficiency one will be red. A commission working group will determine the energy classes and the specific products that must be labelled.

Energy Performance Certificate (EPC)

The purpose of EPCs is to use typical energy-demand patterns for existing similar developments to estimate a likely energy-demand pattern for the prospective development. Therefore, by resorting to a consistent dataset of assumed demand characteristics, an EPC affords prospective purchasers or tenants the opportunity to compare the relative efficiencies of similar properties.

Recommendation reports are also produced with the certificate. These reports are generated by the software tool and list the factors that most affect the outcome of the EPC assessment. Whilst the information within the reports is useful when planning energy-efficiency improvements, it is important to remember that the EPC process relies on a number of assumptions, including assumptions about energy usage. Designing systems to achieve actual efficiencies also benefits from knowledge about the actual consumption.

EPCs are now mandatory for almost all property transactions in the UK and whilst the penalties for not having an assessment are usually relatively small in the context of the overall transaction, the absence of a certificate can potentially cause transactions to stall, or even fall through.

Display Energy Certificates

The Energy Performance of Buildings (Certificates and Inspections) (England and Wales) Regulations 2007 (the 'Energy Performance Regulations') implement

Article 7 of the EU directive in England and Wales. It requires DECs to be displayed prominently in all publicly accessible and public-authority-occupied buildings in England and Wales with an area of more than 1,000 square metres as of 1 October 2008. The approach is seen as a flagship approach to highlight awareness of consumption and the annual progress being made. There are a number of other countries and municipalities that are looking to implement such an approach.

A DEC gives the operational rating of a building, standardised so that one building can be easily compared to another building of a similar type. The features of the certificate are (see Figure 4.1):

- Energy-performance operational rating;
- Total carbon dioxide emissions;
- Previous operational ratings (last three accounting periods);
- Technical and administrative information;
- Accompanying recommendation report with recommendations for improving energy performance (fabric and services);
- Certificate valid for one year and advisory report valid for seven years.

A DEC displays the actual energy performance of a building during operation compared to EPCs, which measure the theoretical energy use. The rating is based on the metered energy used by the building in the previous 12 months, which is compared to established benchmarks to produce an efficiency rating. DECs are produced using the Operational Rating Calculation (ORCalc).

DECs are produced by energy assessors who are members of an accreditation scheme and have appropriate qualifications and competences. A list of accredited assessors with their contact details is held by each individual accreditation scheme. DECs and their associated reports are recorded in a central database, the Non-Domestic Energy Performance Certificate Register.

EU Waste Directive

The directive will clarify a number of definitions that have previously led to uncertainty and variation since the original directive in the 1970s. In particular, definitions for the following are clarified:

- When waste ceases to be waste;
- new definitions for recovery and disposal;
- the introduction of a definition for recycling.

The Waste Framework Directive (WFD; Directive 2006/12/EC) contains the definition of waste. This definition is used to establish whether a material is a waste or not[15]. The WFD also requires member states of the EU to establish both a network of disposal facilities and competent authorities with responsibility for issuing waste-management authorisations and licences. Member states may also introduce regulations that specify which waste-recovery operations and businesses are exempt from the licensing regimes, and the conditions for those exemptions.

An important objective of the WFD is to ensure the recovery of waste or its disposal without endangering human health and the environment. Great emphasis is also placed on the prevention, reduction, reuse and recycling of waste. All construction and demolition wastes are classified as waste under the definition of the WFD and are subject to its requirements through the relevant legislation.

In December 2008, the new WFD (Directive 2008/98/EC) came into force, amending some articles of the current WFD. Amongst others, changes that will come into place include:

- The setting of recycling targets for non-hazardous construction and demolition waste (70 per cent by 2020);
- A provision that would enable the European Commission to adopt EU-wide end-of-waste criteria for specified wastes; a waste specified in this way would cease to be waste when it has undergone a recovery operation and complies with the criteria set by the commission;
- The obligation for member states to set up waste-prevention plans within five years from the adoption of the directive.

The implication for the refurbishment industry will be to promote the reuse and recycling of materials to remove them from the waste stream and thus avoid the significant costs associated with their management and disposal. As a recycled material, this can be sold to produce a revenue stream and, therefore, a secondary income from the project can be identified.

United States

Unlike Europe, the US promotes legislation for energy and sustainability at the state level and, therefore, there are relatively few federal regulations. The American Recovery and Reinvestment Act, more commonly known as the federal stimulus programme, offers greater support for the cost-effective energy reductions achieved through building retro-its. Executive Order 13423 and the Energy Independence and Security Act required a 3 per cent greenhouse-gas reduction to be met annually through to 2015, relative to a 2003 baseline, not only for buildings, but also for space occupied by federal agencies. Refurbishments were expected to meet a 20 per cent reduction relative to the baseline.

The proposed Green Act, known technically as HR2336, could lead to cash incentives for green development. It will set new benchmarks for green building and sustainability for properties that receive financial assistance from the federal Department of Housing and Urban Development.

More than 40 states require targets for sustainability and energy efficiency to be met within their real-estate portfolio. Increasingly, this approach is being applied to the private sector for both new build and refurbishment. California's Green Building Standards Code requires the reduction of water and energy use in buildings through landscaping, appliance efficiency, and the use of green-building-design principles and recycled materials. In addition, California adopted AB1103, similar to Europe's EPBD, in 2007, which became effective from 1 January 2010. Similar energy-disclosure requirements are in various legislative stages in a number of states and cities, including Seattle, New York City and the District of Columbia.

China's Energy Conservation Law

As the fastest-growing economic power, China has set in place a number of measures to reduce its climate-change impacts. The Energy Conservation Law was adopted on 1 November 1997, with amendments in 2007 and 2008. The Act aims to strengthen energy conservation, particularly for key energy-using entities, promote rational utilisation of energy and advancement of energy-conservation technology. Each state is required to implement a system to achieve targets for energy evaluation and energy conservation as part of their overall local-government activities.

Canada

Canadian regulations combine building regulations with incentives, provided at the provincial and municipal level. In December 2008, Toronto adopted its own Green Standard, which requires minimum energy-efficiency and sustainability standards to be met for residential and commercial construction. Properties that meet the more stringent 'Tier 2' criteria also receive performance benefits, including a reduction in development fees[16].

Individual provinces, such as British Columbia and Ontario, have set criteria for energy efficiency to be met, together with state-wide financial incentives – see the box in Section 4.4.

4.4 Financial Incentives and Taxes

On a global basis, direct financial support has been provided to support construction, not only for new build, but also to encourage energy efficiency in the refurbishment of domestic and commercial buildings. As we have seen in the part dealing with market drivers, there is little in place to incentivise or bring about full sustainable refurbishment of buildings, with many governments historically focusing on energy and carbon in the short term.

- The economic downturn has had a major impact on housebuilding across most of Europe, and the Netherlands is no exception, where €320 million will be spent to encourage companies and individuals to invest in energy-efficiency measures for their properties. Homeowners will also be able to invest in energy-saving measures at the lowest possible cost. To achieve this, central government plans to set up a special energy-conservation fund to guarantee low-interest-rate loans taken out by private individuals from institutions in the private sector. The new scheme will help to safeguard employment and provide incentives to homeowners to improve the energy efficiency of their homes.
- The 2005 Energy Policy Act in the US has a provision to extend a tax relief for energy-efficient commercial buildings achieving a certain threshold, which is applicable between 1 January 2006 and 2013. A tax relief of US$1.80 per square foot is available for existing building owners who achieve this threshold from their retro-fit solutions.
- The US Environmental Protection Agency and the Department of Energy have formed an action group to help states achieve the maximum cost-effective,

energy-efficiency improvements possible in offices, buildings, industries and homes by 2020. The Obama administration announced the launch of the State Energy Efficiency (SEE) Action Network in February 2010. To guide its efforts, the group plans to work from the framework set by the 'National Action Plan for Energy Efficiency, Vision for 2025', which was laid out in 2006.

In January, the National Governors Association Center for Best Practices selected six states – Colorado, Hawaii, Massachusetts, North Carolina, Utah and Wisconsin – to participate in the organisation's Policy Academy on State Building Efficiency Retrofit Programs. The academy, funded by the DOE, is designed to help states develop strategies and action plans to improve the energy efficiency of existing buildings and reduce costs and emissions.

Most energy-efficiency initiatives have focused on new construction or on the low-income sector only, often ignoring the substantial energy savings available by retro-fitting existing buildings. The Policy Academy will help states realise energy savings across the board, through comprehensive building-retro-fit programmes.

Canada's ecoENERGY Programme[17]

Through the government of Canada's $3.6-billion ecoENERGY, the initiatives have successfully laid the foundation for many of the services implemented in both new and existing buildings across Canada. The energy-saving initiatives can help reduce annual energy consumption and costs by an average of 20 per cent in existing buildings.

The ecoENERGY Retrofit Incentive for Buildings helps Canadians reduce the payback period of their energy-efficiency projects and increase their return on investment. Commercial and institutional buildings with an area of up to 20,000 square metres (215,279 square feet) are eligible for the incentive. The programme encourages the implementation of multiple and proven retro-fit measures, such as improvements to lighting, heating and cooling systems, as well as to building envelopes. A shorter payback period for some of these measures can help compensate for the longer payback period of others, making it possible for savings to be invested in future retro-fit projects.

The retro-fit incentive is based on the lowest of three amounts: $10 per gigajoule (277.8 kilowatt hours) of estimated annual energy savings; 25 per cent of eligible project costs; or $50,000 per project ($250,000 per organisation). The estimated payback period of the investment needs to be at least one year after taking into account similar incentives from other sources. The application process involves arranging for a pre-project energy assessment before an application is submitted. The average approval period is six to eight weeks, and, after written approval is received, the project can be started and eligible costs can be incurred.

Furthermore, the Office of Energy Efficiency is consulting with provincial governments and other stakeholders to continue developing a voluntary rating-and-labelling system for existing Canadian buildings. This system will help building owners compare the energy performance of their commercial and institutional buildings with similar facilities in their region or across Canada. According to an NRCan 2007 online survey, 86 per cent of respondents expressed support for the labelling of commercial and institutional buildings in Canada and thought that these labels should serve as a benchmarking tool for comparison with similar buildings.

The ecoENERGY for Renewable Heat programme increases the amount of renewable thermal energy used and created, and contributes to cleaner air, by using alternative fuels to heat space and water. The Government of Canada offers up to $80,000 to those in the industrial, commercial and institutional sectors who install active, energy-efficient, solar, air-heating or water-heating systems in existing facilities.

Feed-In Tariff (FiT)

FiTs typically include three key provisions[18]:

- guaranteed grid access;
- long-term contracts for the electricity produced;
- purchase prices that tend towards grid parity over time.

Under a feed-in tariff, eligible renewable-electricity generators (which can include homeowners and businesses) are paid a premium price for any renewable electricity they produce. Typically, regional or national electric-grid utilities are obliged to take the electricity and pay those who generate the electricity.

Different tariff rates are typically set for different renewable-energy technologies, linked to the cost of resource development in each case. The cost-based prices therefore enable a diversity of projects (wind, solar, etc.) to be developed whilst investors can obtain a reasonable return on renewable-energy investments because the purchase prices are based on the cost of generation. This principle was first explained in Germany's 2000 RES Act:

'The compensation rates . . . have been determined by means of scientific studies, subject to the provision that the rates identified should make it possible for an installation – when managed efficiently – to be operated cost-effectively, based on the use of state-of-the-art technology and depending on the renewable energy sources naturally available in a given geographical environment.' (RES Act 2000, Explanatory Memorandum A.)[19]

As a result, the rate is likely to differ amongst the various sources of renewable-energy generation, but also potentially by the size of the project and the location where the project is installed. The rates are designed to ratchet downward over time to track technological change and overall cost reductions. This is consistent with keeping the payment levels in line with actual generation costs over time.

In addition, FITs typically offer a guaranteed purchase for electricity generated from renewable-energy sources within long-term (15–25-year) contracts. These contracts are usually offered in a non-discriminatory way to all interested producers of renewable electricity.

Feed-in-tariff policies have been enacted in 63 countries, across Europe, Australia, parts of Africa, South America and Asia, together with a dozen states in the US.

In 2008, a detailed analysis by the European Commission concluded that 'well-adapted feed-in tariff regimes are generally the most efficient and effective support schemes for promoting renewable electricity'. This conclusion has been supported by a number of recent analyses, including that by the International Energy Agency[20].

In 1990, Germany adopted its 'Stromeinspeisungsgesetz' (StrEG), or its 'Law on Feeding Electricity into the Grid'. The StrEG required utilities to purchase electricity generated from renewable-energy sources at prices that were determined as a percentage of the prevailing retail price of electricity. The percentage offered to solar and wind power was set at 90 per cent of the residential-electricity price, whilst other technologies, such as hydro-power and biomass sources, were offered percentages ranging from 65 to 80 per cent. A project cap of 5 MW was included. Whilst Germany's StrEG (Feed-in Law) was insufficient to encourage costlier sources of generation, such as solar photovoltaics, it proved relatively effective at encouraging lower-cost technologies, such as wind power, leading to the deployment of 4,400 MW of new wind capacity between 1991 and 1999, representing approximately one-third of the global capacity at the time. The StrEG stipulated that renewable-electricity producers would be guaranteed grid access. Similar percentage-based feed-in laws were adopted in Spain, as well as in Denmark in the 1990s[21].

Germany's Feed-in Law underwent a major restructuring in the year 2000, being re-framed as the Act on Granting Priority to Renewable Energy Sources ('Erneuerbare Energien Gesetz', German Renewable Energy Act). In its new form, it has proved to be the world's most effective policy framework for accelerating the deployment of renewable-energy technologies. The new tariff made a number of important changes to its previous policy: 1) the purchase prices were methodologically based on the cost of generation from renewable-energy sources. This led to different prices for wind power, solar power, biomass and biogas sources, and geothermal energy, as well as different prices for projects of different sizes, to account for economies of scale; 2) purchase guarantees were extended for a period of 20 years; 3) utilities were now allowed to participate; and, finally, 4) the rates offered were designed to decline annually based on expected cost reductions, in a mechanism known as 'tariff degression'[22].

Since it has been the most successful, the German policy (amended in 2004 and 2008) often provides the benchmark against which other feed-in-tariff policies are considered. Following the German approach, a number of countries have begun adopting feed-in-tariff policies. These long-term contracts for electricity are typically offered in a non-discriminatory manner to all producers of electricity generated from renewable-energy sources.

Feed-in-tariff policies typically target a rate of return ranging from 5 to 10 per cent. Feed-in tariffs have been associated with a large growth in solar power in Spain and Germany, and in wind power in Denmark, from which these countries now boast the supply of 9, 5 and 20 per cent of their electricity respectively. These systems involve fixed, per-kWh payments that are guaranteed for periods ranging from 10 to 25 years. Feed-in tariffs can be used to accelerate the pace at which renewable-energy technologies become cost-competitive with electricity provided from the grid. The rapid deployment of renewable energy under feed-in tariffs seen in countries like Germany, Denmark and Spain has undoubtedly contributed to reducing technology costs.

Feed-in Tariff in the UK

In the UK, all generation and export tariffs will be linked to the Retail Price Index (RPI), and FITs' income for domestic properties, generating electricity mainly for their own use, will not be taxable income for the purposes of income tax. Once an installation has been allocated a generation tariff, that tariff remains fixed (though it will alter with inflation) for the life of that installation or for the life of the tariff, whichever is the shorter. The FIT provides a guaranteed return of 5 to 8 per cent on the installation of a range of electricity-renewable technologies, effectively providing a secondary revenue stream.

There are a number of opportunities available for existing buildings for the landlord or owner to utilise the tariff:

- Property owners can utilise roof space to provide green electricity to meet corporate commitments; the mechanism can be managed through an energy-services company;
- Roof space or land can be leased to a third party to invest and deliver the FITs, receiving green electricity in return for rentals.

Carbon Reduction Commitment Energy Efficiency Scheme – CRC (UK)

The CRC is a mandatory carbon tax that aims to improve energy efficiency and reduce the amount of carbon dioxide emitted in the UK. It came into effect in April 2010, and has led to a number of other similar taxes in California and Tokyo.

The CRC is now applicable to almost 3,000 organisations and covers most businesses with an electricity bill in excess of £500,000 per annum, including government departments. The intention is for carbon reduction to be met through cost savings, reduced carbon footprints and improved employee awareness.

There are four key drivers for the CRC:

1. Requirement for an organisation's board member to sign off the formal submission of data and reports; this ensures that the legislation has transparency at the most senior level of an organisation;
2. Reputational risks through the production of a league table that will be delineated into peer groups, thereby allowing analysis of comparison for the market;
3. Cost of carbon has been set as an initial £12-per-tonne tax with an expectation for increases to be set on a long-term escalator to further incentivise behavioural change;
4. Requirement to collect accurate data for annual reports through the use of direct metering and validation of the energy consumption; from accurate information comes the ability to set targets for reduction in key areas where over-consumption is occurring.

There are lessons to be learnt from the implementation of the CRC, which are being reviewed by a number of other municipal bodies and countries prior to confirming implementation. These include:

* Organisational complexity was underestimated, particularly with the large number of multi-national firms operating with shared-ownership subsidiaries; mergers and acquisitions; and selling of business units;
* League table, providing the transparency criteria, needs to be detailed enough to identify where organisations have made real reductions and changes to their emissions, taking into account business changes, including organic growth, divestments, mergers or reduction in properties; the current approach based on absolute emissions penalises those organisations who are growing, and benefits those with offshore emissions outside the UK;
* Alignment with other carbon instruments, in particular, the EU Emissions Trading Scheme, and other financial incentives being promoted.

RICS Red Book

A two-tier property market based on energy performance is anticipated, supported in part by changes made in April 2010 to the RICS Red Book. The book will link property values to the sustainability of buildings. From then on, the principle of environmental performance commanding increased value will be 'common currency'.

Initiatives such as the Carbon Reduction Commitment Energy Efficiency Scheme (CRC) would start to feed through to property-rental values and valuations by quantifying the cost of carbon. This provides one method of helping assess the impact of sustainability on valuation. The likely net effect of sustainability on a building's value will vary widely depending on its type, location and individual circumstances.

Carbon Tax (US)

Tax risks further complicate the problem as building investments extend beyond the foreseeable future of carbon taxation. Current legislative action at the federal level includes Representative John B. Larson's carbon-pricing bill, American's Energy Security Trust Fund Act of 2009. Representative Larson's basic premise is to establish 'a national carbon tax rate of $15 per ton of carbon dioxide in 2012 . . . that would rise annually by $10/ton, with an alternate annual increment rate of $15/ton if required to meet emission reduction targets to be pre-established by U.S. EPA'[23].

Current legislation targets these taxes at the high polluters such as oil refineries, coal facilities and import sites. Individual building owners are not directly impacted by such taxation, yet. However, numerous state and local governments have either taxes or incentive programmes geared toward carbon reduction that do directly impact organisations.

For example, the State of California Employee Trip Reduction programme taxes employers that have high-volume traffic generation from their commuting employees. Maricopa County, Arizona, provides a similar incentive to reduce traffic volume. And the State of New Jersey is piloting a volunteer programme to establish the same behaviour. Whilst the stages of development and target audiences vary, the intent is clear. Owners will pay to pollute; the unknown extent is simply a matter of timing and degree of penalty. When owners have large portfolios that cross multiple geographies, and potentially, therefore, multiple local and state legislatures, both the complexity and the risk increase. Portfolio reinvestments offer owners the opportunity to optimise more by reusing existing building stock, thereby reducing initial capital expenditures and costs over time.

Liability

Executives and their companies are being held liable for activities that contribute to global warming. In the SEC Disclosure Requirements, from early 2010, companies must weigh the impact of climate change when reporting risks to their investors. For the first time, companies need to track, analyse, and report such things as energy use and efficiency, GHG emissions, and other aspects of their business, of which the reporting has until now been a purely political exercise. Directors and officers (D&O) insurance protects companies from the actions of its directors and officers. Without it, entire corporations would be at risk from the actions of their executives. It's a key part of the corporate structure, and it has now entered a grey area due to climate change.

Several cases have already been brought to the Supreme Court as groups target executives and their corporations for their activities that contribute to global warming. At the centre of this debate is a common inclusion in D&O that excludes 'pollution' from coverage. The question is: 'Do greenhouse gases constitute pollution?' If they do, then executives are not covered by corporate insurance and the company may be exposed to the risk of litigation. The EPA issued a recent ruling that GHG is a pollutant, and this sets a precedent that could adversely impact corporations and place them on the wrong side of this debate.

For corporate officers, it's all about protecting the company and shareholder value, which includes avoiding the risk of litigation at all costs. It appears that – even without any climate legislation in place – the question of executive liability is forcing the issue of climate change in the American corporation today.

Similar requirements across Europe apply, linked to the EU Company Law Directive, which is driven by, and related to, the US regulations.

4.5   Climate Change

Sustainable refurbishment is essential if we are to meet our national targets for reducing carbon emissions. About 150 million tonnes of carbon dioxide are released each year from the housing stock, with older buildings contributing disproportionately. The contribution of best-practice, sustainable refurbishment in bringing this level down by over 60 per cent by 2050, the IPCC recommended target, is central and a key part in the market transformation of the sector.

Kyoto Protocol

In June 1992, the governments of the world signed the UN Framework Convention on Climate Change (UNFCCC) to combat the growing threat from climate change. Its ultimate objective was 'to achieve . . . stabilisation of greenhouse gas concentrations in the atmosphere at a level which would prevent dangerous anthropogenic interference with the climate system'[24]. Subsequently, the Kyoto Protocol was adopted by 159 countries in 1997 in Kyoto, Japan. Among the 117 countries that have ratified the Protocol since, 40 of them are industrialised nations that agreed to reduce their emissions of six key greenhouse gases by 5.2 per cent from their 1990 levels by 2012. The United States bailed out of the Kyoto Protocol in 2001, with President Bush saying that the treaty would place too heavy a burden on the US economy.

Kyoto came into force in February 2005 as a binding agreement when Russia ratified the treaty; based on each country's emissions in 1990, the total number of countries signed up to the Kyoto Protocol had emissions above 55 per cent of the total amount. It has been ratified by almost all of Europe, Japan, Canada, Russia and New Zealand, collectively accounting for 61.4 per cent of the required total. The United States (36.2 per cent) is the only developed nation not to ratify.

In 2005, the European Union launched an emissions' trading system, under which European companies that emit less carbon dioxide than allowed can sell unused allotments to those who overshoot the target. The motive for profit was expected to drive efforts and technology and bring 'substantial cuts' in emissions of carbon dioxide. Globally, the new carbon-trading market is valued at around $142 billion per year[25].

Commitments to the Kyoto Protocol could be met through the purchasing of carbon credits, without reducing emissions of greenhouse gases at all. Since the 1990 baseline for Kyoto obligations, many smokestacks in the former Soviet Union have gone cold, enabling Russia and the other successor countries to sell their carbon credits for the consequent drop in emissions. Forest-sink credits arise under a Kyoto provision that recognises that plantation forests established on land not previously forested, while they are growing, are withdrawing carbon dioxide from the atmosphere. New Zealand expects to have more forest-sink credits than it would need to cover the 33 per cent increase in its emissions since 1990.

The Kyoto Protocol defines several mechanisms that are designed to allow developed countries to meet their emission-reduction commitments (caps) with reduced economic impact (IPCC, 2007), including the International Emissions Trading (IET)[26]. Under the treaty, for the 5-year compliance period from 2008 until 2012, nations that emit less than their quota will be able to sell assigned-amount units to nations that exceed their quota[27]. It is also possible for developed countries to sponsor carbon projects that reduce greenhouse-gas emissions in other countries. These projects generate tradable carbon credits that can be used in meeting their caps. The project-based Kyoto Mechanisms are the Clean Development Mechanism (CDM) and Joint Implementation (JI). The CDM covers projects taking place in non-Annex-I countries, whilst JI covers projects taking place in Annex-I countries. CDM projects are supposed to generate 'real' and 'additional' emission savings, that is, savings that are attributable only to the specific CDM project. The question as to whether or not these emission savings are genuine is, however, difficult to prove[28].

Post-Kyoto Discussions

One of Kyoto's limitations is that it does not place any constraints on developing countries, whose collective emissions of greenhouse gases are estimated to overtake those of developed countries within 20 years. The rationale is that the accumulated emissions of industrialised countries over the past 150 years are doing the damage and that rich countries are better placed to incur the costs of reducing emissions than are countries still struggling to escape from poverty.

The Kyoto Protocol can only be a 'first step' towards negotiating deeper cuts in greenhouse-gas emissions, primarily because it sets very low targets compared to those that scientists say are necessary in order to keep climate change under control. Recent rounds of international climate talks have already included negotiations on greenhouse-gas emissions beyond 2012 to include many developing countries that are currently excluded. However, divisions exist over whether to extend the Kyoto Protocol beyond 2012, or to move fully to an alternative pledge-and-review system for the reduction of emissions. Climate-control finance for the developing nations has been made available: $30 billion of fast-start finance offered for 2010–12, which will be stepped up to the annual $100 billion promised from 2020.

Emissions' Trading Schemes

Emissions' trading is a market-based approach used to control pollution by providing economic incentives for achieving reductions in the emissions of pollutants. It is a form of carbon pricing. Typically, a government body sets a limit, or cap, on the amount of carbon that can be emitted. The limit, or cap, is allocated, or sold, to firms in the form of emissions' permits, which represent the right to emit or discharge a specific volume of carbon. Firms are required to hold a number of permits equivalent to their emissions. The total number of permits cannot exceed the cap, limiting total emissions to that level. Firms that need to increase their emissions' permits must buy permits from those who require fewer permits.

The transfer of permits is referred to as a trade. In effect, the buyer is paying a charge for polluting, whilst the seller is being rewarded for having reduced emissions. Thus, in theory, those who can reduce emissions most cheaply will do so, achieving the pollution reduction at the lowest cost to society.

The overall goal of an emissions' trading plan is to minimise the cost of meeting a set emissions' target. The cap is an enforceable limit on emissions that is usually lowered over time, aiming towards a national target for emissions' reduction. In other systems, a portion of all traded credits must be retired, causing a net reduction in emissions each time a trade occurs. In many cap-and-trade systems, organisations that do not pollute may also participate, thus environmental groups can purchase and retire allowances or credits and hence drive up the price of the remainder according to the law of demand. Corporations can also prematurely retire allowances by donating them to a non-profit entity and then be eligible for a tax relief[29].

For trading purposes, one allowance, or CER, is considered equivalent to one metric tonne of carbon dioxide emissions. These allowances can be sold privately or in the international market at the prevailing market price. The allowances are traded and settled internationally and hence may be transferred between countries. Each international transfer is validated by the UNFCCC. Each transfer of ownership within the European Union is additionally validated by the European Commission.

Climate exchanges have been established to provide a spot market in allowances, as well as a futures-and-options market to help discover a market price and maintain liquidity. Carbon prices are normally quoted in euros per tonne of carbon dioxide. Currently, there are six exchanges that are trading in carbon allowances: the Chicago Climate Exchange, NASDAQ OMX Commodities Europe, European Climate Exchange, PowerNext, Commodity Exchange Bratislava, and the European Energy Exchange. Managing emissions is one of the fastest-growing segments globally with a market estimated to be worth about €30 billion in 2007.

In 2003 the **New South Wales (NSW) state government** (in Australia) unilaterally established the NSW Greenhouse Gas Abatement Scheme to reduce emissions by requiring electricity generators and large consumers to purchase NSW Greenhouse Abatement Certificates (NGACs)[30]. This prompted the roll-out of free energy-efficient, compact, fluorescent light bulbs and other energy-efficiency measures, funded by the credits. The government announced that a cap-and-trade emissions' trading scheme would be introduced in 2010, however, this scheme has since been delayed until 2013[31].

The **New Zealand Emissions Trading Scheme** (NZ ETS) is a national all-sectors all-greenhouse-gas uncapped emissions' trading scheme first legislated in September 2008 by the government and amended in November 2009[32]. Tradable emission units can be issued by free allocation to emitters, with no auctions in the short term. A transition period is operating from 1 July 2010 until 31 December 2012. During this period the price of New Zealand Emissions Units (NZUs) will be capped at NZ$25. Also, only one unit will need to be surrendered for every two tonnes of carbon dioxide equivalent emissions, effectively reducing the carbon price to NZ$12.50 per tonne[33]. The Climate Change Response Act 2002 aims to reduce emissions from business-as-usual levels and to fulfil New Zealand's obligations under the Kyoto Protocol.

The **European Union Emission Trading Scheme (EU ETS)** is the largest multinational, greenhouse-gas emissions' trading scheme in the world. It is one of the EU's central-policy instruments to meet its cap set in the Kyoto Protocol, designed to be cost-effective. Phase I commenced operation in January 2005 with all 15 (now 25 of the 27) member states of the EU participating, as well as Norway, Iceland, and Lichtenstein who joined in January 2008. The scheme operates by the allocation and trading of greenhouse-gas emissions' allowances throughout the EU – one allowance represents one tonne of carbon dioxide equivalent. The programme caps the amount of carbon dioxide that can be emitted from large installations with a net heat supply in excess of 20 MW, such as power plants and carbon-intensive factories, and covers almost half (46 per cent) of the EU's carbon dioxide emissions.

Phase I permits participants to trade amongst themselves and in validated credits from the developing world through Kyoto's CDM. During Phases I and II, allowances for emissions have typically been given free to firms, which has resulted in windfall profits for them[34]. A number of design flaws have limited the effectiveness of the scheme. In the initial 2005–07 period, emission caps were not tight enough to produce a significant reduction in emissions. The total allocation of allowances turned out to exceed actual emissions. This drove the carbon price down to zero in 2007. This oversupply reflects the difficulty in predicting future emissions, which is necessary in setting a cap.

Phase II saw some tightening, but the use of JI and CDM offsets was allowed, with the result that no reductions in the EU will be required to meet the Phase II cap. For Phase II, the cap is expected to result in an emissions' reduction in 2010 of about 2.4 per cent compared to expected emissions without the cap (business-as-usual emissions). For Phase III (2013–17), the EU has proposed a number of changes, including:

• Setting an overall EU cap, with allowances then allocated to EU members;
• Tighter limits on the use of offsets;
• Unlimited banking of allowances between Phases II and III;
• A move from allowances to auctioning.

Tokyo, Japan: Originally, Japan had its own cap-and-trade system that had been in place for some years, but was not effective[35]. Japan has its own emissions' reduction policy – this climate-control strategy is enforced and overseen by the Tokyo Metropolitan Government. The first phase, which is similar to Japan's original scheme, runs until 2014, and organisations will have to cut their carbon emissions by 6 per cent; those who fail to operate within their emissions' cap will from 2011 be required to purchase emissions' allowances to cover any excess emissions, or to invest in renewable-energy certificates or offset credits issued by smaller businesses or branch offices. Firms that fail to comply will face fines. The long-term aim is to cut carbon emissions in the metropolis by 25 per cent from 2000 levels by 2020[36].

United States

In 2003, New York State came up with a proposal and secured commitments from nine north-east states to form a cap-and-trade carbon dioxide emissions' programme for power generators, called the Regional Greenhouse Gas Initiative (RGGI). This programme was launched on 1 January 2009 with the aim to reduce the carbon 'budget' of each state's electricity-generation sector to 10 per cent below their 2009 allowances by 2018[37]. In 2006, the California Legislature passed the California Global Warming Solutions Act, AB-32, which was signed into law by Governor Arnold Schwarzenegger. Thus far, flexible mechanisms, in the form of project-based offsets, have been suggested for five main project types including building energy. However, a recent ruling from Judge Ernest H. Goldsmith of San Francisco's Superior Court states that the rules governing California's cap-and-trade system were adopted without a proper analysis of alternative methods to reduce greenhouse-gas emissions[38]. Since February 2007, seven US states and four Canadian provinces have joined together to create the Western Climate Initiative, a regional greenhouse-gas emissions' trading system. In July 2010, a meeting took place to further outline the cap-and-trade system, which, if accepted, would curb greenhouse-gas emissions by January 2012[39].

The 2010 US federal budget proposes to support clean-energy development with a 10-year investment of US$15 billion per year, generated from the sale of greenhouse-gas (GHG) emissions' credits. Under the proposed cap-and-trade programme, all GHG emissions' credits would be auctioned off, generating an estimated US$78.7 billion in additional revenue in FY 2012, steadily increasing to US$83 billion by FY 2019. The American Clean Energy and Security Act (H.R.

2454), a greenhouse-gas cap-and-trade bill, was passed on 26 June 2009, in the House of Representatives. It was never passed in the Senate. The big Republican wins in the November 2010 US Congressional election have further reduced the chances of a climate-control bill being adopted during President Barack Obama's first term.

Renewable energy certificates, or 'green tags', are transferable rights for renewable energy within some American states. A renewable-energy provider is issued with one green tag for each 1,000 kWh of energy it produces. The energy is sold into the electrical grid, and the certificates can be sold on the open market for profit. They are purchased by firms or individuals in order to identify that a portion of their energy is from renewable sources, and this is voluntary. They are typically used as an offsetting scheme or to show corporate responsibility, although their issuance is unregulated, with no national registry to ensure that there is no double-counting. However, it is one way that an organisation could purchase its energy from a local provider that uses fossil fuels, but back the purchase with a certificate that supports a specific wind-power or hydro-power project.

Cost-Effective Emissions' Reductions

In its 2007 report, the IPCC identified a series of sectors that held the most promise for achieving large-scale and cost-effective emissions' reductions. The panel determined that cost-effective emissions' reductions in building could achieve 30 per cent against the estimated 2020 baseline. This would be the equivalent to 3.2 gigatonnes of carbon dioxide[40]. These reductions would come from reduced demand for electricity and reduced fuel based on the following themes:

1. Improved building insulation;
2. Heating and cooling efficiencies;
3. Energy-efficient lighting;
4. Reduced plug Loads from energy-efficient appliances and business machines.

The potential reductions would be dependent on existing proven technologies, and would not require experimental, or untested, technologies to be applied. Using discount rates of between 3 per cent and 10 per cent, the IPCC calculated that the emissions' reductions could be cost-negative – that is, the cost of implementation would be less than the savings realised[41]. The measures would therefore produce a positive return on investment.

4.6   Corporate Responsibility

Evidence in the current market shows that whilst aspects of sustainable refurbishment that are not related to energy directly are not captured within typical projects' costs and benefits, many organisations are moving ahead on bespoke requirements aligned with their brand values.

Companies are under increasing pressure from key stakeholders to be transparent about their values, principles and performance regarding sustainable development. One response to this pressure is the increase in 'sustainability reporting'. It is quite clear, however, that reporting is only the tip of the iceberg. Companies will find it difficult to continue to produce relevant and reliable reports without having internal management-and-information systems that support this undertaking. The key challenge is to integrate sustainable-development issues into mainstream business processes and systems to determine how well companies put their policies into practice.

The role of organisations in the global economy is increasingly coming under the scrutiny of investors, analysts and pressure groups. Activities such as the riots, as seen at G8 summits, are no longer greeted with public scorn, but with questions about why governments and private organisations are not pulling their weight to support a sustainable global economy (see Figure 4.2).

Further drivers to encourage the collection of data to improve transparency and disclosure include the Carbon Disclosure Project[42], which acts as an intermediary between shareholders and corporate companies on climate-change-related issues, by providing primary climate-change data from the world's largest businesses. The data is obtained from responses to CDP's annual information request sent on behalf of institutional investors and purchasing organisations. Pension funds and other institutional investors are actively assessing the impacts that sustainability will have on their portfolio values. As part of their shareholding, they encourage businesses to integrate the sustainability risks into their business strategy and risk-management approach.

There are a number of underlying principles for organisations to deliver sustainability:

- Sustainability must be the organising principle incorporated as part of the central business plan and processes;
- Natural resources and systems have finite limits and all economic activity must be constrained within those limits;
- Economics must ensure that basic needs are met and maintained equally across the globe;
- Cost of pollution must be internalised – captured as part of the life-cycle-costing and decision-making process;
- There is no blueprint – off-the-shelf solutions will not be appropriate or applicable to align with the business strategy;
- Lack of scientific certainty should not be used as a reason to delay taking cost-effective measures to prevent damage.

There is a recognition by many organisations that corporate responsibility can improve their brand value and help to grow the business. The properties that these organisations choose to occupy also need to reflect the brand values they are promoting. Recent examples include GE, M&S, Phillips, Siemens, and Tata, all seen as positive examples of organisations that have strong corporate-responsibility brands that have been specifically developed to differentiate these organisations in their respective marketplaces. Each of these organisations has also recently developed new, or refurbished, properties that are flagship sustainability-focused premises.

As an increasing number of organisations are moving towards a brand image involving promoting corporate responsibility, there will be a corresponding increase in the number of sustainable buildings and refurbishments necessary to maintain this profile. Construction and developer organisations in this field have an additional incentive to ensure that their properties meet certain sustainability

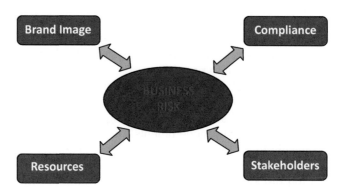

Figure 4.2: Relationship between business risks and activities which CSR and SD support.

standards. Organisations, including British Land, Len Lease and Skanska, seek to ensure that their prime stock meets high-sustainability criteria.

Similarly, many governments are pushing for occupation of high-specification sustainable buildings. The Dutch government are seeking to occupy 'C'-rated buildings and above, through the EPBD. In Germany, the demand is driven by tenants who require sustainable buildings to meet their corporate responsibilities.

The approach of Corporate Responsibility (CR) incorporates a greater emphasis on global activities and responsibility to those in the Third World through the procurement or manufacturing of goods. CR is also broader in its context, taking into account elements of health and safety, ethical responsibility and investment in personnel.

Whilst critics of CR have argued that it is not part of core business – that is, to do business and make a profit – many companies no longer agree. The market is changing, and CR is becoming a vital part of staying competitive, retaining talented staff, and satisfying customers' expectations. Brand-name companies like Shell, Nike and Nestle have discovered through high-profile scandals concerning the environment, human rights, health and labour conditions that they have to take society's concerns seriously in order to preserve their licence to operate. Increased scrutiny via the internet also means that organisations have to be rigorous in the way in which they address these issues. 'Greenwashing', or covering up their activities with symbolic gestures, will only have limited effectiveness before stakeholders, including employees, demand greater transparency and more action. The next wave will see CR consolidated and integrated as a core-business strategy. Those companies being left behind will be at a distinct disadvantage in the coming decade as CR becomes firmly embedded worldwide.

In response to the increasing awareness of CR and pressures by the investor community, the number and importance of specialised, ethical investment funds has increased significantly. In January 2002, 279 green, social and ethical funds operated in Europe. From January 2000 until the second quarter of 2001, ethical financial products available to private clientele had increased by 58 per cent [43]. Moreover, the global demand for ethical investing opportunities is growing. Many property firms form part of these ethical investment trusts together with major occupiers, aiding the drive for increased sustainability in buildings.

Whilst the concept of corporate responsibility has embedded itself within multinationals, the culture and processes of small-and-medium-size enterprises (SMEs) has hardly changed at all – especially for those at the smaller end of the spectrum. Small-business owners show little awareness and interest due to a lack of resources, skilled staff and technical expertise, poor access to finance, and they often fall victim to burdensome red tape and regulation. Whilst this leads to SMEs being a significant source of environmental pollution, given the number and range of such organisations, many have integrated links with the local community and with their cultural and religious environment, thus providing a strong social focus for employment. The role of larger organisations is to support the SMEs and, through the supply chain, build the skills and standards for the smaller businesses and local communities. Responsible companies can help by engaging with local businesses and supporting the development of small or micro enterprises whose services they can purchase.

Supply Chain

The involvement of the supply chain with an organisation's brand, subject to their own sustainability performance, is increasing, with some high-profile cases pushing the agenda further. Some organisations, such as BT and Gap, have robust programmes in place.

Walmart Require Suppliers' Focus on Sustainability

Walmart has shifted gears to focus more on store upgrades than expansions, to reinvigorate its private brands, and to get undeniably serious about sustainability. These challenges are not going away. However, Walmart is trimming the number of its suppliers and focusing on building 'strategic relationships' with companies, large and small, that demonstrate leadership. Some companies supplying Walmart today may not be supplying Walmart tomorrow. Fortunately, there are numerous bright spots – with sustainability at the epicentre.

In 2009 Walmart introduced its supplier-sustainability assessment. As suppliers processed the assessment and began to understand how it would impact their organisations, certain recurring themes arose in the form of perceived barriers – reasons that suppliers give for not embracing sustainability. All of these are rooted in fair and realistic thinking, but might also lead to inaction. And Walmart is turning away from suppliers who offer excuses instead of innovation. The company would prefer to build strategic, future relationships with suppliers that look at challenges such as sustainability as an opportunity to improve their businesses, reduce costs, and bring the slogan, 'Save Money, Live Better', to life for the customer.

4.7   Skills

A significant problem with being able to retro-fit buildings in a sustainable manner rests with the level and competence of those within the profession. Over the past few decades the focus has been on new buildings with turnkey products to fit within the building. A growing attention to off-site manufacturing has contributed to reducing the level of skills necessary within the industry – or certainly to fragment it further into specialised roles.

The advent of refurbishment as a major part of the construction landscape, and increasingly that of sustainable refurbishment, has required a different set of skills to understand the existing building and its impacts prior to adding a further layer. Failure to link these elements together in order to understand how the building will operate as a whole can lead to unusable buildings, at worst, or to a significant waste of money, at best.

There are variances dependent on the level of refurbishment required. Certainly, smaller-scale refurbishments (levels 1 and 2), requiring cosmetic or minor changes in the building, have arguably taken place consistently and, therefore, a competent workforce is in place. Likewise, the major refurbishment at level 5 is effectively building from the shell again, with skills available from the fit-out contractors. It is for the middle levels of 3 and 4, when significant parts of a building are being refurbished, combined with some elements of the existing building to remain, that skills' deficiencies do exist.

For both the lower and higher levels, with minor changes and rebuilding from the shell, there is anticipated to be an increase in the number of jobs globally where re-training or upskilling of the workforce is likely to be required. The Green Deal in the UK focuses on improving energy efficiency within domestic and commercial buildings through measures such as greater insulation and increased air tightness. It is anticipated that this programme will provide over 100,000 jobs in the short term, building up to more than 200,000 jobs during the peak. Similar levels are seen in South Africa, where the refurbishment programme is anticipated to generate more than 800,000 jobs over a seven-year period focusing on improving housing[44].

There are a number of gaps that exist predominantly in the upstream design and financial activities, and also in the highly technical areas. Key gaps that have been identified include:

- Institutional limitations, particularly integrated legislation and financial mechanisms;
- Technological barriers and the need to import green technology in many cases;
- Human capital and skills' shortages, largely in upstream production and financial activities;
- Lack, and immaturity, of key organisations, particularly ESCOs for supporting, facilitating and for information awareness.

Governments are stepping in to provide financial support in this area to promote the development of skill-sets.

Early in 2010, LA City Council passed a 'Green Building Retrofit Ordinance' to retro-fit all city-owned buildings with an area larger than 7,500 square feet, or built before 1978, with a target of hitting LEED Silver-level certification. The green jobs' element of the project involves assigning the highest priority to retro-fitting buildings that are located in low-income communities, as well as to buildings that directly benefit those same communities, such as libraries and recreation centres. The ordinance was developed by the LA Apollo Alliance, a broad coalition of community, labour and environmental groups. It will put people to work in green jobs, generate revenue for local businesses, save LA taxpayers up to $6 million in energy costs and cut global-warming pollution.

Amongst the other goals of the ordinance are:

- To establish a pipeline to green careers by recruiting disadvantaged workers into the city training programmes that can train and connect unemployed and underemployed workers from under-served communities to construction apprenticeship positions on green retro-fits and to job placement elsewhere in both the public and the private sector;
- To foster inner-city, economic development by supporting local minority and women-owned, green business development;
- To ensure that quality, green products are being used and purchased locally, to encourage local, green manufacturing and the purchase of locally produced green goods that prevents waste and prohibits toxic chemicals that are unhealthy for workers in the retro-fitting;
- To foster public-sector career development within the city by hiring city workers from city training programmes, and by upgrading part-time workers to full time.

Failure to tackle some of these skills' shortages can lead to additional barriers in the take-up of technologies and practices. A public-sector body in London included combined heat-and-power technologies as part of refurbishing a number of premises. The lack of skilled experience in commissioning the units would have led to the removal of the CHP systems, without the involvement of engineers from Germany who were required to correctly commission the units.

Endnotes

1 Emerging Trends in Real Estate 2011, Urban Land Institute and PWC
2 4th Annual Sustainability Survey conducted by CoreNet Global and Jones Lang LaSalle, February 2011
3 Oxford Institute for Sustainable Development (OISD) at Oxford Brookes University, conducted as part of the Investment Property Forum Research Programme
4 Mercer LLP, Energy efficiency and real estate: opportunities for investors, published by Ceres 2010
5 Unlisted Real Estate Funds – Environmental Review, Aviva Investors and Environment Agency, 2008
6 Doing Well By Doing Good?, Piet Eichholtz and Nils Kok, Maastricht University and John Quigley, University of California, Berkeley; RICS Research March 2009
7 Existing Buildings Survival Strategies, Arup and Davis Langdon, June 2009
8 Pivo, Gary and Fischer, Jeffrey D., Investment Returns from Responsible Property Investments: Energy Efficient, Transit-orientated and Urban Regeneration Office Properties in the US from 1998 to 2007, October 2008 (revised March 2009)
9 Fuerst, Franz and McAllister, Patrick; Green Noise or Green Value? Measuring the Price Effects of Environmental Certification in the Commercial Buildings, School of Real Estate and Planning, Henley Business School, 25 April 2009
10 Nils Kok, Piet Eichholtz, Rob Bauer, Paulo Peneda: Environmental Performance: A Global Perspective on Commercial Real Estate, The European Centre for Corporate Engagement (ECCE)
11 Energy Efficiency Retrofits for Commercial and Public Buildings, Pike Research, 2010
12 Nils Kok, Maastricht University, PRI Workshop, January 2009

13 Is sustainability reflected in commercial property prices: an analysis of the evidence base; C-SCAIPE, School of Surveying & Planning, Kingston University: Sarah Sayce, Anna Sundberg, Billy Clements; January 2010

14 Directive 2002/91/EC on the Energy Performance of Buildings, 16 December 2002, http://eur-lex.europa.eu/LexUriServ.do?uri=OJ:L:2003:001:0065:0071:EN:pdf

15 http://aggregain.wrap.org.uk/waste_management_regulations/background/european.html

16 Toronto Green Building Standard; www.toronto.ca/planning/greendevelopment.htm

17 Natural Resources Canada website at www.nrcan-rncan.gc.ca

18 Mendonça, M. (2007). Feed-in Tariffs: Accelerating the Deployment of Renewable Energy, London: EarthScan

19 Germany, Renewable Energy Sources Act (RES Act) (2000), 'Act on Granting Priority to Renewable Energy Sources', Federal Ministry for the Environment, Nature Conservation and Nuclear Safety (BMU), Accessed 15 May 2009 at: http://www.wind-works.org/FeedLaws/Germany/GermanEEG2000.pdf

20 International Energy Agency (IEA) (2008), Deploying Renewables: Principles for Effective Policies, ISBN 978-92-64-04220-9

21 Germany, Stromeinspeisungsgesetz (StrEG) (1990), 'Germany's Act on Feeding Renewable Energies into the Grid of 7 December 1990', Federal Law Gazette I, p. 2663, unofficial translation, Accessed 9 July 2009 at: http://wind-works.org/FeedLaws/Germany/ARTsDE.html

22 Jacobsson, S.and Lauber, V. (2006), 'The Politics and Policy of Energy System Transformation – explaining the German Diffusion of Renewable Energy Technology', Energy Policy (34); pp. 256–276

23 http://www.larson.house.gov/index.php?option=com_content&task=view&id=798&Itemid=1

24 UN Framework Convention on Climate Change – http://unfccc.int/

25 http://www.reuters.com/article/2011/07/11/us-carbon-schemes-idUSTRE76A2GJ20110711

26 Glossary J-P. In (book section): Annex I. In: Climate Change 2007: Mitigation, Contribution of Working Group III to the Fourth Assessment Report of the Intergovernmental Panel on Climate Change (B. Metz et al. eds.), Cambridge University Press, http://www.ipcc.ch/publications_and_data/ar4/wg3/en/annex1sglossary-j-p.html

27 'UNFCCC Countries 1990 to 2012 emissions targets'. UNFCCC website. 2008-05-14 http://unfccc.int/kyoto_protocol/background/items/3145.php

28 'World Development Report 2010: Development and Climate Change' pp. 265–267, The International Bank for Reconstruction and Development, The World Bank, http://go.worldbank.org/BKLQ9DSDU0

29 Sullivan, Arthur and Steven M. Sheffrin. Economics: Principles in action, Upper Saddle River, NJ, 2003, ISBN 0130630853

30 'The Greenhouse Gas Reduction Scheme'. NSW: Greenhouse Gas Reduction Scheme Administrator, 2010-01-04, http://greenhousegas.nsw.gov.au/

31 Department of Climate Change and Energy Efficiency (2010-05-05). 'Carbon Pollution Reduction Scheme'. Press release. http://www.climatechange.gov.au/en/media/whats-new/cprs-delayed.aspx

32 'Climate Change Response (Emissions Trading) Amendment Act 2008 No 85', www.legislation.govt.nz, Parliamentary Counsel Office, 2008-09-25, http://www.legislation.govt.nz/act/public/2008/0085/latest/DLM1130932.html

33 MfE (September 2009), 'Summary of the proposed changes to the NZ ETS'. Emissions trading bulletin No 11, Ministry for the Environment (MfE), NZ Government. http://www.mfe.govt.nz/publications/climate/emissions-trading-bulletin-11/index.html#summary

34 CCC (December 2008), 'Chapter 4: Carbon markets and carbon prices. In: Building a low-carbon economy – The UK's contribution to tackling climate change, The First Report of the Committee on Climate Change, December 2008' http://www.theccc.org.uk/reports/building-a-low-carbon-economy, p. 149

35 'Tokyo emissions trading plan may become a model for others', World Council for Sustainable Development http://www.wbcsd.org/plugins/DocSearch/details.asp?type=DocDet&ObjectId=Mzc0MzU

36 Business Green (2010-04-08). 'Tokyo kicks off carbon trading scheme', The Guardian http://www.guardian.co.uk/environment/2010/apr/08/tokyo-carbon-trading-scheme

37 Memorandum of Understanding – Regional Greenhouse Gas Initiative

38 Barringer, Felicity (4 February 2011), 'California Law to Curb Greenhouse Gases Faces a Legal Hurdle', The New York Times http://www.nytimes.com/2011/02/05/science/earth/05emit.html

39 California, New Mexico and 3 Canadian provinces outline regional cap-and-trade program, Los Angeles Times, 28 July 2010

40 B.S. Fisher et al., 'Issues Related to Mitigation in the Long Term Context' in Climate Change 2007: Mitigation. Contribution to Working Group III to the Fourth Assessment Report of the Intergovernmental panel on Climate Change (Cambridge, UK; Cambridge University Press 2007)

41 G.J. Levermore, 'A Review of the IPCC Assessment Report Four Part 1: The IPCC Process and Greenhouse Gas Emission Trends from Buildings Worldwide', Building Services Engineering Research and Technology vol. 29, no. 4, pp. 349–361

42 www.cdproject.net/responding-companies.asp

43 Sustainable Investment Research International, January 2002

44 Green Jobs Creation Through Sustainable Refurbishment in the Developing Countries, Ramin Keivani, Joe Tah, Esra Kurul, Fonbeyin Abanda; International Labour Organisation, 2010

5 Developing a Business Plan and Strategy

For owners, the decision as to whether to refurbish or redevelop will primarily depend on the commercial viability of the options available and the need to maximise the economic performance of a building for both owner and occupier. Weighing up the costs, risks and benefits, by means of well-established development-appraisal techniques, will help to determine the financial viability of a project. The cost of doing nothing is likely to far outweigh the cost of change or the penalty for doing nothing.

However, included in the assessment must be the non-financial activities such as occupant performance, which can help to improve the letting of a building; increase the performance of staff; and make the business a more attractive proposition. If a refurbishment scheme is able to produce workspace that contributes positively to the performance of its occupiers while delivering a satisfactory level of return to the investor, it should be viewed as the most economically sustainable option.

Chapter Learning Guide

This chapter will review the criteria necessary to develop an effective strategy, this being the key element for a successful refurbishment.

- Cost-saving benefits and risks associated with refurbishment projects;
- Existing barriers to take projects forward and their implications;
- Factors to improve the sustainability profile of the development;
- Development of a cohesive strategy to pull the various strands together.

Sustainable Refurbishment, First Edition. Sunil Shah.
© 2012 Sunil Shah. Published 2012 by Blackwell Publishing Ltd.

> Key messages include:
>
> - Whilst refurbishment is typically less costly than a new build, the increased risks need to be evaluated effectively and managed through the process;
> - A number of improvements can be made to the scheme to appeal more to the end-user at the strategy stage.

5.1   Costs and Risks to Refurbish

It is often assumed that refurbishment is a low-cost alternative to redevelopment principally at the medium-scale and major-scale level (see Figure 1.1). However, the costs of refurbishment can be influenced by a number of factors that will vary greatly from project to project. More consideration should be given by property owners as to how the characteristics of existing buildings can be exploited to increase the financial viability of refurbishment. The potential for such savings should be fully assessed and factored into development-appraisal models.

Key differences between refurbishment and redevelopment where cost savings might be made include[1]:

- **Planning and legal costs** – Refurbishment works are likely to make faster progress through the planning system, and building-regulations requirements might be less rigid than with new build. In addition, the burden of developer contributions might be reduced or avoided.
- **Demolition costs** – Demolition and waste-disposal costs will, in most cases, be lower in a refurbishment project due to the reuse of building materials. An associated benefit is the reduction in waste, resulting in savings from disposal costs and landfill tax.
- **Building-material costs** – Lower overall building-material costs can be achieved through the retention and recycling of existing building materials. The preservation of architectural features might also enhance value to potential occupiers.
- **Maintenance of income** – A phased refurbishment might mean that parts of an existing building could remain in occupation while works are carried out, providing an ongoing income for the owner. Refurbishment is often quicker to complete than redevelopment, thus reducing the void period.
- **Tax relief** – Property-tax relief in a variety of forms offers a range of financial advantages. Refurbishments can offer one of the highest levels of tax relief available.

Alongside potential cost savings are some substantial risks and technological barriers to overcome, each with potentially significant cost implications. Again, each case is unique, but there are common areas of refurbishment risk including[2]:

- **Existing structure** – Constraints arise from the condition of the existing building fabric, form and orientation, including the unpredictability of the effects of demolition and temporary works on the retained fabric. Physical constraints include slab-to-slab and ceiling-to-floor height limitations, limited riser capacity, and plantroom space. The redistribution of services and means of escape might be problematic. There is likely to be a need to improve the performance of the building fabric to provide improved thermal insulation and control of glare, while optimising the use of natural light in office space.

- **Contingency requirement** – A higher level of contingency might be required for the increased risk of unforeseen costs associated with refurbishment work, and mechanisms will be required to deal with any unexpected difficulties.
- **Safety issues** – Consideration should be given to the possibility of the unexpected occurrence of hazardous materials such as asbestos, and to the possibility of complex planning and sequencing of the construction programme, which might require expert risk assessment and management.
- **Procurement** – The heightened risks and technical challenges of refurbishment mean that contractors with specialist expertise might be required, further raising costs associated with the procurement process.

5.2   Barriers to Refurbishment

Whilst there are many benefits and drivers promoting the sustainable refurbishment of properties, there is still relatively limited activity in this area, primarily due to a series of barriers, whether perceived or actual.

- Financial considerations – building owners traditionally focus upon initial capital costs, without considering the relevant medium-term and long-term cost and benefits. If owners utilise a short-term investment time frame or do not accurately determine the overall rate of return, the higher, upfront costs of more energy-efficient equipment discourage their purchase and use.
- Disconnection between costs and benefits – if building owners bear the costs of sustainability measures and if the benefits, expressed as lower energy costs, accrue only to tenants, there might be a disincentive to invest. If the benefit cannot be recouped by the cost bearer, the investment will not occur.
- Lack of knowledge and lack of experienced workforce – a lack of practical understanding amongst building owners about energy efficiency and sustainable buildings, including the owners' overestimates of the first cost premium, is widespread. A WBCSD report highlighted a contradiction, whereby even though there existed a high awareness of green buildings amongst real-estate professionals and building owners, less than 13 per cent of those surveyed had been involved in a green building. The lack of experienced service providers ultimately raises the cost of sustainable buildings through higher risk levels attributed to costs[3].
- Increase in risk and uncertainty – the lack of knowledge about practical, sustainable measures amongst real-estate professionals compounds the problem of assessing sustainable buildings and energy-efficient practices. Because some measures rely on building practices that are perceived as being new to a marketplace that is traditionally slow to adapt, there is uncertainty about both physical and financial performance. In addition, sustainable-building practices might require the use of new suppliers and contractors, which could increase

the risk profile for the application and lower the risk-adjusted financial returns from the retro-fit.

- Ignoring small opportunities for sustainable improvements – many small measures are overlooked by building owners, especially in residential structures. At the aggregate level, improvements from such measures are substantial, but if the ROI for these measures does not satisfy owners' minimum return thresholds, the measures will not be undertaken.

Although some of these impediments have, in the past, held back the implementation of sustainable buildings, there is a growing consensus that these barriers are being removed. As property owners, tenants, contractors and bankers begin to appreciate the value proposition of a sustainable strategy, there is less resistance to, and greater comfort with, a sustainable refurbishment.

5.3 Delivering Commercially Viable Refurbishment

Deciding on the most economically sustainable path requires a balancing of the costs, risks and benefits of carrying out the work. At the initial concept and design phases of a project, a key consideration should be that to achieve economic sustainability, the finished product must benefit the economic performance of the occupiers as well as that of the owners. The design must facilitate the occupier's drive to maximise productivity through high levels of workspace efficiency. Office space should be created with a high level of in-built flexibility with regard to space planning, IT infrastructure and services, allowing optimal occupational densities to be achieved.

When assessing the commercial viability of a potential refurbishment project, it is important to understand the costs, risks and benefits to the investor over the projected holding period. In order to achieve this, life-cycle costing should be incorporated into the standard development-appraisal process to evaluate the various development options available. Life-cycle costing involves estimating the present value of the total cost of the proposed development project over its entire operating life (including initial capital cost, occupation costs, operating costs and the cost or benefit of the eventual disposal at the end of its life).

The key variables for consideration when comparing the life-cycle costs of refurbishment with redevelopment will include the extent of construction works, the discount rate applied, tax allowances, the proposed building life and the nature and condition of the existing building.

There is no one-size-fits-all strategy as so much depends on a particular building's existing structure. The first stage, when considering the options available, should be to undertake an assessment of the performance of the existing building. Particular attention should be given to energy use, occupier satisfaction, operational efficiency, the condition of internal fittings and the external building fabric. Comparing the results against benchmarks will highlight areas that require improvements to be made. Traditional development-appraisal techniques, incorporating life-cycle-costing methods, can then be used to estimate the commercial viability of a range of potential options for the building, from basic refurbishment through to complete redevelopment.

5.4   Factors to Consider to Improve the Performance

One of the principal advantages of refurbishment over new build is the potential to reduce resource consumption through the reuse of structural elements, building materials and services. However, the existing structure can be a burden, placing constraints on what can be achieved.

Identifying features of existing buildings that can be exploited should be one of the first considerations when assessing the refurbishment potential. Each case will be unique, but there are several common areas worthy of consideration to significantly enhance a comprehensive asset strategy that incorporates tenant demand, regulatory risk and competitiveness:

Market-specific criteria:
- Market and sub-market conditions, including rents, occupancy rates, rental concessions; and pipeline sector-specific new space available in the market (planned or under construction);
- Tenant demand for sustainable space, capturing the tenant interest and acceptable cost premium for sustainable space and as part of this, capturing the specific interest in sustainable buildings;
- Regulations (current and pending) that might affect building use, value and competitiveness; Tenant lease structure (e.g. gross, net) for building tenants together with options for passing the cost of building improvements through to tenants.

Property-specific criteria:
- Indoor environmental quality and occupant comfort, capturing the issues of thermal comfort, ventilation and lighting conditions for building occupants;
- Building-energy consumption, use profile and energy intensity for the building; comparing the energy profile with comparable and similar buildings to identify the gaps and opportunities to improve any performance difference;
- Building-water consumption and waste-water discharge and use profile for the building; comparing the water profile with comparable and similar buildings to identify the gaps and opportunities to improve any performance difference;
- Facility maintenance and operations to understand the condition of the existing plant and equipment to determine whether anything requires repair or replacement in the short term;
- Site and environmental impacts to promote soft-landscaping and hard-landscaping features.

Sustainability-specific criteria:
- Embodied energy is conserved during a refurbishment project compared to that of a new build due to the reduction in the materials necessary and reuse of the facade and structure;
- Location and orientation, when considered early, can mitigate some of the burden placed on mechanical services to control the internal environment, such as the consideration of passive-design strategies;
- Reuse of building materials and land to improve the sustainable rating of the building.

5.5   Defining the Strategy

There is no one-size-fits-all refurbishment strategy, as it depends more on the owners' objectives and, where leased, on the desire for asset competitiveness and asset value. Owner-occupied buildings do require a different approach, with a broader view encompassing financial performance to maximise the productivity of employees or to enhance to corporate-responsibility profile. In this way, owner-occupiers will assess financial returns by including both direct costs and indirect financial benefits that contribute to the organisations' performance and brand value.

The identification of the property performance, deficiencies and specific retro-fit opportunities will help to define the forward refurbishment strategy. Critically, the constraints must also be understood in terms of capital limitations, existing leases or time limits that might remove the potential for certain objectives.

The timing and duration of the retro-fit are influenced by pending regulations, tax and funding provisions, and the competitive market. The expense of a refurbishment project might create a need to implement and complete the renovation in a relatively short time frame. For owner-occupiers, or for those with long-term, stable tenancy, a longer-term view can be taken, which could deliver a more thorough sustainability plan.

The budget for the refurbishment project should encompass the ability to meet the desired outcome, providing an enhanced leasing, tenant/employee retention, reduced operating cost and improved employee performance. The time frame of the refurbishment project will have a significant impact on the size of the budget. High- cost replacement items, such as plant equipment, will have a cost impact, but they are necessary items for the building to meet its objectives. An effective way of managing the cost is to align the refurbishment project with the time frames when capital expenditure is required, such as for the replacement of the plant equipment at the end of its practicable life.

Endnotes

1 UK Offices: Refurbishment vs redevelopment; GVA Grimley; Winter 2010
2 ibid
3 WBCSD, 8 September 2008

6 Managing Delivery

Some of the activities that have the most uncontrolled environmental impact are projects varying from minor refurbishment and churn, through to building extensions. Importantly, sustainability considerations should be reviewed within the design, construction and operational phases of a project in order to minimise the impact. The reasoning behind reviewing the impacts at each phase is not only to provide mitigation of the known risks, but also to identify new risks that become apparent from the additional information provided. Moreover, the inherent nature of projects means that they constantly change and, therefore, the initial design might not be the final chosen design and this could produce different risks.

It is important that relevant legislation and corporate, trade or industry standards are incorporated into the process. This provides the basis for ensuring that the environmental impacts can then be managed and minimised throughout the duration of the project. To facilitate this, sustainability should be included as an agenda item within meetings at every stage, with particular focus at the design-concept, construction and completion stages to identify and incorporate environmental best practice.

Sustainable Refurbishment, First Edition. Sunil Shah.
© 2012 Sunil Shah. Published 2012 by Blackwell Publishing Ltd.

Chapter Learning Guide

This chapter will identify the processes required to deliver the refurbishment:

- Framework built around the concepts of the brief, design, construction and operation to form the delivery structure;
- Key issues to focus on, as applicable to small-scale refurbishments or, alternatively, as applicable to major changes;
- Sustainability issues affecting the project and how they can be managed.

Key messages include:

- Early identification of the sustainability baseline and the goals that will help to identify how this can be bridged;
- Sustainability-performance indicators help to focus attention on designing in the necessary measures, but these must be communicated to the occupiers in order to obtain the best operational response..

6.1   Delivering a Sustainable Refurbishment

An approach that is aligned with the construction process is one that divides the refurbishment process into four phases: brief, design, construction and operation. The key is to transfer the knowledge through each cycle and learn lessons from specific feedback from the operational stage all the way back to the briefing stage. This forms part of a continuous improvement loop (see Figure 6.1).

1. Brief – there should be a commitment from key stakeholders to deliver an environmentally sustainable building. This vision should be set out and fully incorporated into the development brief. The performance of the existing building should be established, which will enable the setting of targets for sustainability improvement. A dedicated budget should be set for the sustainability elements with a 'champion' appointed to ensure that the original vision is maintained. An experienced design team should be selected with the power to implement solutions to achieve sustainability.

2. Design – reducing energy demand, whilst minimising water use, and providing biodiversity, transport and recycling measures should all be key considerations. A wide range of options should be explored, utilising both sustainability-data analysis and life-cycle-costing appraisals. Budgets should be carefully managed and the final design approved and signed off by the client. Emissions' targets should be included in procurement arrangements for the construction phase.

3. Construction – effective project management to focus on sustainability and careful selection of contractors with experience in sustainable refurbishment.

Figure 6.1: The sustainability cycle.

Site workers should be encouraged to understand the importance of energy efficiency with the construction process being consistently monitored against objectives. Energy-monitoring equipment, built into the project, will allow the tracking of performance within the completed building.

4. Operation – in the period following refurbishment, knowledge of the improvements made should be conveyed to the occupants of the building. Building energy-management systems should be carefully designed and those responsible for the operation of the building should understand how to correctly operate new plant, systems and controls. Post-occupancy evaluations will help to ensure that the improved systems are functioning correctly. In-use performance measures should be monitored with the results used to highlight potential areas for further improvements. Additionally, details of the sustainability features of the refurbished building should be clearly communicated to stakeholders.

6.2   Minor Refurbishment Approach

Chapter 1 outlined a series of five different levels of refurbishment. The mechanism to implement each project is different dependent on the level of change taking place. The lower levels of change are more likely to occur with the occupants remaining within the building and therefore the focus will be on buy-in, communication and gaining support. The higher levels of refurbishment are associated with major changes when the building is not occupiedso the changes can be more structured against the environmental targets to be met.

1. Ensure Buy-in

Even the most state-of-the-art eco fit-out can fail if the buy-in of the occupants hasn't been ensured. Without the commitment and enthusiasm of the people who will be using the building on a day-to-day basis, many of the features that make a truly sustainable building just don't happen. You can enforce certain procedural practices, for instance, auto-shutdown of PCs and monitors at night, but it's better for everyone if the move to sustainability feels like a choice, rather than an imposition.

2. Green Design Is Beautiful

Eco office design means stylish and modern furnishings and furniture that have been manufactured with a consideration of how their being affects the planet. It means light and airy spaces where people want to work. Build sustainability into the design of your new space from the very start, and it won't cost you much more than a 'standard' office.

3. Think Natural

A major feature of the sustainable office is the amount of natural light available. This doesn't necessarily require a relocation to somewhere with floor-to-ceiling windows in order to be achieved (if you're lucky enough to have them already, though, then make the best use of them). Look at your floor plan – if the removal of a wall means that natural light will be able to spread deeper into the office, then consider it. You can use glazing to partition space, and maintain a sense of privacy if necessary, whilst still allowing light to penetrate.

4. Make an Investment

An office fit-out is the perfect opportunity to replace worn-out, inefficient air-conditioning and heating systems. Consider replacing them with energy-efficient alternatives – the initial investment might be higher, but running costs are likely to be lower. And some equipment can be eligible for tax rebates.

5. Get Rated

It does require some consideration early on in the process, but getting an evaluation against an environmental-rating system, such as BREEAM or LEED, has the benefit that its criteria can help specify what to include within the office fit-out. Even if a formal rating is not necessary, the framework and approach provide a good guide as to what makes a sustainable space.

6. During Refurbishment

During the construction phase, there are many decisions that will reduce the overall environmental impact of the site-construction activities. For instance, monitoring the consumption of water, gas and electricity on site against set targets will help to reduce wastage. Other standards to measure are how much waste from site is recycled, how much local labour is used (more commuter miles generate more carbon dioxide emissions), and how much carbon dioxide is emitted by delivery vehicles. It should also be ensured that all timber used on site is certified by the Forest Stewardship Council, or similar, to confirm that it has been sourced sustainably.

The above items provide a guide to the sort of things you can do to make your office more environmentally friendly. However, it's not a case of all or nothing – you can decide the level of commitment you want to make, whether it's to the fullest extent or just the preference of choosing sustainable products.

Key Points

- Ensure the buy-in of the people who use the space;
- Eco design can be stylish and modern;
- Maximise natural light for an office in which people will be keen to work;
- Invest in energy-efficient equipment, and reap the benefits by saving money on bills;
- Use environmental-rating systems to your advantage;
- Construction provides a key opportunity to make green choices.

6.3 Major and Comprehensive Refurbishment Approach

Where a more significant change within a building is being made, a greater opportunity is available to assess building assets, setting appropriate targets and identifying key upgrade initiatives to make sure that sound investment decisions are being made.

The approach develops the brief-design-construction-operation model described in Part 6.1 into a framework to determine the necessary level of sustainability inclusion. The approach supports the identification and assessment of sustainability opportunities, targeted at the initial briefing stage, as the project brief is conceived and developed. Design responses to sustainability requirements are continuously monitored and developed throughout the design, construction and, where relevant, operation stages. To address requirements at brief, design, construction and operation stages, a 12-step approach has been developed to support the implementation of sustainability, as indicated in Figure 6.2.

The twelve steps outlined in the framework are:

Step 1: Define the Baseline: The first stage is to assess the current performance of the building including the condition, utilisation and occupancy satisfaction to help determine the baseline. It will also identify areas for improvement in the existing building to achieve what the market and tenants/employees are expecting, which will help to formulate a series of options.

Step 2: Options' Assessment: Once the baseline has been developed, a series of options can be formulated to determine the next step and finalise the targets and goals to be achieved. A great deal of thought needs to go into this stage before any works take place to determine the costs and benefits.

Step 3: Sustainability Project Brief: With a clear direction of the goals to be achieved and the targets to be met, a briefing document is created to accompany the wider design brief. This document will also allow sign-off to the riskier

	1.	Define the Baseline
Brief	2.	Options Assessment
	3.	Sustainability Project Brief
Design	4.	Implementation into design
	5.	Measure progress
	6.	Sustainability Performance Indicators
Construct	7.	Sustainable Construction
	8.	Handover & Commissioning
	9.	Defects and performance
Operate	10.	Operation – feedback to design
	11.	Education and Culture
	12.	Occupancy Evaluation

Figure 6.2: Framework to deliver sustainable refurbishment.

design stage with an understanding of what is to be achieved together with an outline cost. Its purpose is therefore to aid decision making.

Step 4: Implementation into Design: The first stage in the design process is to translate the sustainability project brief into the concept design.

Step 5: Measure Progress: Through the design process, from concept into more detailed engineering solutions, the implementation of sustainability measures should be tracked throughout to ensure that they are being effectively incorporated. The use of modelling, in particular, Building Information Modelling (BIM), has helped the understanding of how the engineering and sustainability elements can benefit each other – for example, the insulation benefits of a green roof.

Step 6: Sustainability Performance Indicators: Throughout the design process, thought should be given to defining the performance measures that will determine the success of the project from a sustainability perspective.

Step 7: Sustainable Construction: The choice of a suitable construction partner is key to the success of the project and, therefore, tender documentation must include questions that will enable the tendering contractors to demonstrate their understanding of the project. Clauses within the contract to support the achievement of the sustainability in line with performance indicators should be included.

Step 8: Handover and Commissioning: Effective handover to the operators of the building, whether existing or new, is an ongoing process involving the building operators at the earlier design stage through to post-commissioning dialogue. At this stage, the building will be seasonally commissioned and handed over based on how the building will be operated when occupied.

Step 9: Defects and Performance: Over the first two years following handover, a close dialogue with the designers and project team should be maintained to highlight any defects arising and to realign the building operations as necessary to ensure that the systems are effectively bedded in. This approach will also provide the link between the behavioural-change aspects and the measurement of occupancy performance.

Step 10: Operation: Feedback to design: with the project having operated for a number of months, feedback on the design is critical to better understand how well the design has met the occupancy requirements and where improvements can be made. For owners of multiple buildings with a common culture, this is key to enable the creation of better designs to be suitable for the way in which the business functions. Reassessment of the sustainability-performance indicators will help to demonstrate the financial and non-financial benefits that have been achieved.

Step 11: Education and Culture: The culture change for the business is a critical element that will ensure that the technical changes implemented within the design are successful. The occupants of the building will need to buy in to the changes and, therefore, the education programme starts at the beginning in order to make individuals aware of their role.

Step 12: Occupancy Evaluation: The final activity is the use of occupancy evaluations to understand how successful the refurbishment has proved to be for the occupants in a working situation and to help tweak the environmental controls and layout as necessary.

6.4   Green Leases

Most leases do not promote environmental or social improvements, stating fixed energy and waste costs as part of the service charge. Sustainability issues should be incorporated within the lease agreement through the negotiation of a 'green lease' where appropriate. Whilst this may be requested as something of a wish list, which not all premises may be able to provide, it does enable informed decisions to be made. The increase of such requests will also help to drive the 'green market' forward – as more tenants request such lease terms, the market will be encouraged to provide the measures described below.

Key points about a green lease are detailed below:

- It should be appended to the main lease so that if there is a non-compliance with the green-lease activities, it will not impact on, and cause a breach of, the main lease;
- There is no standardised approach to green leases, which creates some flexibility to develop a mechanism that works for both parties;
- There need to be incentives for both parties;
- The green lease should cover:
  - Scope of sustainability and the targets;
  - Building rules;
  - Ways to achieve the targets, for example, a green building committee;
  - Dispute resolution;
  - Penalties.
- Effectively, the green lease should cover:
  - What needs to be achieved;
  - How it will be achieved;
  - Over what time frame;
  - Who is responsible;
  - What happens if it goes wrong.

- The green lease should be developed from an environmental-management-system approach.

Increasingly, large corporate organisations are looking to implement green-lease mechanisms within their portfolio. Bank of America, Beacon Capital Partners, Deutsche Bank and its real-estate investment-management arm, RREEF, JPMorgan Chase, Jones Lang LaSalle's investment arm, LaSalle Investment Management, and Whirlpool Corporation have all committed to the Green Lease Action Plan and its guiding principles and have pledged to apply the concepts to properties they own, manage or occupy[1].

The efficient use of energy and other resources in buildings requires joint action from property owners and occupiers, but there are disincentives that often prevent the two sides from investing in the necessary capital improvements. Although studies have shown that market values are higher, and that vacancies and costs are lower, in green commercial buildings, landlords often see the upfront investments and long-payback periods for energy and other efficiency projects as outweighing increased market value. Obtaining financing for green upgrades can be challenging. And tenants, though they derive the benefit of lower utility costs and other advantages, have a tough time making a case for improvements to office space in which they have no long-term vested interest.

The Green Lease Action Plan is trying to create 'win-win scenarios economically and environmentally' for building owners and tenants. Several efforts are under way to help green leasing gain traction:

- **In 2008, BOMA International**, the industry association of building owners and managers, issued its green-lease guide.
- **The Real Property Association of Canada** developed its REALpac Green Office Lease in 2008 as well. The most recent version of that document and their Green Lease Guide for Commercial Office Tenants were released in early 2012.
- **The California Sustainability Alliance** released version 2.0 of its Green Leases Toolkit for landlords and tenants in 2011.
- **Better Buildings Partnership**
- **Australian Green Lease**, backed by the government

The three guiding principles of the initiative are:

1. Landlords and tenants should agree to operate the buildings as sustainably as is commercially viable.
2. The value of energy savings achieved through building-efficiency improvements should be available to pay for the improvements.
3. To whatever extent is feasible, usage of, and demand for, resources throughout the buildings should be measurable and transparent to both the landlord and the tenant.

In committing to the principles, participants agree to:

- Establish green-lease principles to influence owner–occupier agreements and act on these principles across the portfolio over time.

- Require leasing agents who work on behalf of participating organisations to undertake a basic orientation about sustainability, green-lease principles, and ways to remove barriers to sustainability in leases.
- Establish/adopt green site-selection criteria for tenants and consider these criteria for new-space acquisition.
- Establish a standard for landlords to communicate key energy and environmental ratings to tenants and to prospective tenants, and to deploy this process at 50 per cent of their properties within three years.

Some suggestions and questions about helping to develop a green lease are listed below. There is no correct approach other than the one that benefits both the landlord and the tenant.

- Sharing the cost of recycling schemes with other tenants;
- Sharing the cost of energy-efficient equipment with other tenants;
- Including an incentive for energy-saving and recycling initiatives in the calculation of service-charge rates;
- Including a clause for the protection of any environmentally-friendly equipment or building alterations within the dilapidation clauses of an agreement;
- Are noise restrictions imposed to limit the times when works can take place?
- Do any specific trees (or vegetation) have to be maintained as part of the lease?
- What is the payment provision for utilities – gas, electricity and water?
- Is the service charge paid as a percentage or as a flat rate?
- Including a request to define the number of disabled parking spaces;
- Stating that preference is for direct-meter payment, and, if not available, asking what rebates are available to incentivise energy-efficiency returns?
- Requesting details of how waste-disposal costs will be managed and the availability of the relevant reports;
- Stating that information is to be provided about the system specification, controls and zoning, to be able to monitor energy consumption and to have the ability to modify heating and cooling patterns within the building;
- Including a request for staff facilities to be located in the building or nearby, to provide such facilities as a food service, luncheon clubs, health clubs, hotels and retail facilities.

6.5   ISO 14001 Environmental Management Standards

The use of ISO14001 to implement sustainability measures as part of a refurbishment project will provide a robust mechanism to identify the various environmental issues affecting the project and the ways in which these can be proactively managed, controlled and reduced where feasible.

The ISO14001 standard is the most widely recognised and internationally respected EMS and it provides a common approach regardless of country, activity and size. It specifies a framework of control for an environmental-management system against which an organisation can be certified by an external body to a standard that looks for continual improvement through the identification and control of environmental impacts. The standard itself is a short and succinct document written in a legalistic manner; it contains approximately 89 different objectives that must be achieved to the satisfaction of the auditor in order for the applicant to gain the certificate.

ISO14001 forms part of the ISO14000 series – a number of documents focused on providing a standardised methodology for delivering environmental practices within business. The 14000 series includes a range of documents to help organisations to identify, manage, audit and reduce their environmental risks. The text box provides a list of the current documents, which are constantly being reviewed. The system is most commonly used in the UK and Japan, although it is gaining popularity across the globe.

ISO 14000 Series Documentation

- ISO 14004 provides guidance on the development and implementation of environmental-management systems;
- ISO 14010 provides the general principles of environmental auditing (now superseded by ISO 19011);
- ISO 14011 provides specific guidance on the audit of an environmental-management system (now superseded by ISO 19011);
- ISO 14012 provides guidance on the qualification criteria for environmental auditors and lead auditors (now superseded by ISO 19011);
- ISO 14020, 14021 and 14024 provide information on environmental labelling and declaration issues;
- ISO 14030+ provides guidance on performance targets and monitoring within an environmental management system;
- ISO 14040+ covers life-cycle issues and eco-efficiency assessments;
- ISO 14050+ provides environmental terms and definitions, and details of material-flow accounting;
- ISO 14060+ provides standards on greenhouse-gas accounting and verification;
- ISO14063 provides guidelines for environmental communication.

The standard is focused on internal environmental issues, and can be applied to refurbishments to deal with the various sustainability impacts (see Figure 6.3). For each of the impacts identified, mechanisms to control, and, where possible, reduce, the impacts should be implemented, and the benefits should be promoted.

External certification means that organisations can demonstrate to shareholders, regulators and the public that their system has been audited, in the same way as are their financial accounts, by those with appropriate professional skills and knowledge. The information provided by a certified system will be seen as being more credible. Other benefits of external certification include:

- Confidence that the system meets recognised requirements and standards;
- Enhanced value and assurance to customers in the supply chain;
- Independent review of the way in which the organisation is committed to its activities and their associated impacts on the environment;

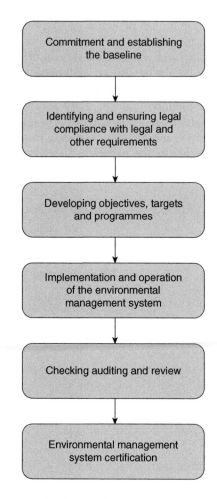

Figure 6.3: Step by step EMS development.

- Closer involvement of employees and other stakeholders;
- Protection of reputational value.

ISO/TS 21931:2006 Sustainability in building construction

This standard provides a framework for the assessment of the environmental performance of construction works. Part 1: Buildings is intended to provide a general framework for improving the quality and comparability of methods for assessing the environmental performance of buildings. It identifies issues to be taken into account when using environmental-assessment methods for new or existing buildings in the design, construction, operation, refurbishment and deconstruction stages. It is not an assessment system in itself but is intended to be used in conjunction with, and following the principles set out in, the ISO 14000 series of standards.

6.6 Energy Star

The well-known Energy Star programme was introduced in 1992 as a voluntary labelling programme designed to promote energy-efficient products. Whilst the first labelled products were computers and monitors, the programme has expanded to include a broad range of items, from major appliances to whole commercial buildings. This trend has been driven not only by the increasing interest in reducing energy consumption and environmental impact, but also by an increasing number of regulations requiring better transparency for consumers. In addition, commercial companies have realised that the Energy Star label is preferred by their customers because it indicates a lower cost of operation.

For building owners, Energy Star has the potential to help managers quickly and easily track their energy history and trends so that they can take concerted steps to reduce energy consumption. Owing to the widespread establishment of regulations requiring building-energy assessment, as well as the incidence of rising energy costs, tracking energy use is a key business practice. From a societal standpoint, this is particularly important as, for example, buildings in the US consume more than 60 per cent of all energy. A proven way to reduce wasted energy through the refurbishment process is to participate in Energy Star. Not only can managers assess and improve their facilities' energy usage with this programme, but they can also save money, reduce their impact on the environment, quantify the impact of operational improvements, and increase their property value.

The primary hurdle is that single-tenant and multi-tenant building managers everywhere have a difficult task in obtaining permission from their tenants to collect energy-usage data, verifying and compiling it for the whole building and providing it to Energy Star. In addition, they have to depend on their utility company to provide usage data. Requesting this data – and waiting for the utility company to collect, vet and deliver this information – has historically been a slow and frustrating process, and this is a process that has to be completed monthly to comply with Energy Star requirements. Many building managers who have encountered this difficulty have opted not to invest the time and resources required to participate in Energy Star. However, for an increasing number of buildings, the programme is no longer optional; new laws now require a commercial Energy Star rating at the time of building sale or rental, and to be eligible for government grants.

In New York City, local law 0476 requires city buildings of more than 10,000 square feet and private buildings of more than 50,000 square feet to track electricity and water usage as from 2011. This represents more than 20,000 buildings – which will place a huge burden on local utilities and building managers to comply with, as they will need to come up with a solution that can be maintained as the programme develops.

In the city of Denver, Colorado, Executive Order 123 requires new construction and major renovations of existing and future city-owned and city-operated buildings to be designed to earn Energy Star certification. This includes being benchmarked in Portfolio Manager, the programme's interactive energy-management tool that allows managers to track and assess energy consumption for an entire commercial building.

6.7 Managing-Delivery Checklist

Table 6.1 provides a summary of the items to be discussed during a sustainable-refurbishment project, capturing where specific items are necessary, recommended or nice-to-have options.

Table 6.1: Managing-delivery checklist

Minimum Standard	Recommended	Nice to have
Implement an environmental-management system for the project and subsequent operation		
Use a sustainability tool to 'test' design and construction proposals for new buildings and refurbishments	Target the top rating of building specification	Conduct post-construction review to check the achievement of the top rating
Specify that the contractor is to be registered with the Considerate Constructors Scheme	Contractor to achieve the highest rating under the Considerate Constructors Scheme	
Determine the impacts of the facility on existing utility and infrastructure availability		
Conduct focused, staged sustainability reviews during the project, to be carried out by an independent agent; develop benchmarks for future use and identify lessons learnt	Performance measures used to incentivise the incorporation of sustainability criteria	
Identify tax incentives available to offset costs for renewable technologies	Provide ring-fenced funds for sustainability initiatives	
Develop a building users' guide/logbook for the end-users to provide a non-technical summary for the operation and maintenance of the facility	Develop a building users' guide/log book in direct consultation with the end-users to ensure that it meets all requirements	
Include FM involvement during pre-commissioning, commissioning (including seasonal) and quality monitoring to ensure that the elements are installed and operational as designed	Provide training to staff and FM on using the new equipment and ensure that O&M manuals are updated	

Endnote

1 http://www.greenprintfoundation.org/Default.aspx

7 Managing Impacts

The need to meet the demands of both regulators and occupiers is encouraging the owners of property to reassess the environmental sustainability of their office portfolios. Although achieving sustainability via refurbishment need be neither difficult nor expensive, the interrelationships between costs, risks and benefits need to be carefully considered.

Identifying sustainability in the existing building stock is problematic, largely because there is no universally accepted assessment method. The focus of current mandatory and voluntary certification schemes is on development and construction rather than on the in-use efficiency of the existing offices. For example, energy-performance certificates relate to the energy-efficiency capability of buildings, not to the actual energy consumption by the occupiers. Voluntary environmental-impact-measurement ratings are now well established, which enable existing buildings to be assessed, but so far their take-up has been slow. Most prominent is the BRE Environmental Assessment Method (BREEAM) and the Leadership in Energy and Environmental Design (LEED).

Although there is a lack of a universally accepted environmental-sustainability rating system, there are common areas of agreement between the principal codes, both mandatory and voluntary. These can be grouped under the broad headings of energy, water, waste and indoor environmental conditions. However, there is a clear need to move to a place where there is agreement and ideally for the creation of a common metric for the measurement of sustainability that allows us to compare building projects.

Sustainable Refurbishment, First Edition. Sunil Shah.
© 2012 Sunil Shah. Published 2012 by Blackwell Publishing Ltd.

Chapter Learning Guide

This chapter will review the standards and methodologies available to minimise the negative, and promote the positive, benefits of sustainability within the design and construction of the refurbishment.

• Overview of international environmental standards and a more detailed review of LEED and BREEAM;
• Benefits of a tailored approach aligned to the company culture.

Key messages include:

• Early engagement and use of sustainability standards can help to identify and implement cost-effective measures;
• Costs to achieve certification can be prohibitive, but might bring additional marketing benefits by making the building easier to sell.

7.1 International Standards

With the increased interest in green building practices, a number of organisations have developed standards, codes and rating systems that enable government regulators, building professionals and consumers to determine green building with confidence. In some cases, codes are written so that local governments can adopt them as by-laws to reduce the local environmental impact of buildings.

Green-building rating systems such as BREEAM and LEED help consumers determine a structure's level of environmental performance. They award credits for optional building features that support green design in categories such as location and maintenance of building site, conservation of water, energy and building materials, and occupant comfort and health. The number of credits generally determines the level of achievement.

Green-building codes and standards, such as the International Code Council's draft International Green Construction Code, are sets of rules created by standards' development organisations that establish minimum requirements for elements of green building such as materials, or heating and cooling.

Table 7.1 shows some of the major building environmental-assessment tools currently in use[1].

IPD Environment Code

Launched in February 2008, the code was developed as a good-practice global standard for measuring the environmental performance of corporate buildings. Its aim is to accurately measure and manage the environmental impacts of corporate buildings and to enable property executives to generate high-quality, comparable performance information about their buildings anywhere in the world. The code covers a wide range of building types (from offices to airports) and aims to inform and support the following:

Table 7.1: Major building environmental-assessment tools

Country	Assessment Tool
Australia	Nabers
	Green Star
Brazil	AQUA
	LEED Brasil
Canada	LEED Canada
	Green Globes
China	GBAS
Finland	PromisE
France	HQE
Germany	DGNB
	CEPHEUS
Hong Kong	HKBEAM
India	Indian Green Building Council (IGBC)
	GRIHA
Indonesia	Green Building Council Indonesia (GBCI)
	Greenship
Italy	Protocollo Itaca
	Green Building Counsil Italia
Japan	CASBEE
Jordan	EDAMA
Korea	KGBC
Malaysia	GBI Malaysia
Mexico	LEED Mexico
Netherlands	BREEAM Netherlands
New Zealand	Green Star NZ
Pakistan	IAPGSA Pakistan Institute of Architecture Pakistan Green Sustainable Architecture
Philippines	BERDE
	Philippine Green Building Council
Portugal	Lider A
Republic of China (Taiwan)	Green Building Label
Singapore	Green Mark
South Africa	Green Star SA
Spain	VERDE
Switzerland	Minergie
United Arab Emirates	Estidama
United Kingdom	BREEAM
United States	LEED
	Living Building Challenge
	Green Globes
	Build it Green
	NAHB NGBS
	International Green Construction Code

- Creating an environmental strategy;
- Inputting to real-estate strategy;
- Communicating a commitment to environmental improvement;
- Creating performance targets;
- Environmental-improvement plans;
- Performance assessment and measurement;
- Life-cycle assessments;
- Acquisition and disposal of buildings;
- Supplier management;
- Information systems and data population;
- Compliance with regulations;
- Team and personal objectives.

IPD estimate that it will take approximately three years to gather significant data to develop a robust set of baseline data that could be used across a typical corporate estate.

7.2   LEED and BREEAM

There are two recognised, global certification schemes in use. BREEAM (Building Research Establishments (BRE) Environmental Assessment Method) was developed in the UK in 1990 and is used widely across Australia, Canada and New Zealand. LEED (Leadership in Energy and Environmental Design) was developed in the late 1990s and has gained credence across the US, Canada and, more recently, in China and India.

Both schemes are excellent methodologies to incorporate environmental criteria into the design process. However, they are best used as a vehicle to encourage greater sustainability practices at the initial briefing stages, rather than as a direct measure of what to include. Traditionally, assessments are performed at the detailed design stage to assess a number of points and to include elements in order to gain the higher rating, if this is easily achievable. Innovation and benefits are limited, so that there could be additional practices, not noted in the methodologies, that would provide performance and cost benefits to the building.

Both schemes offer a range of benefits, from environmental to financial:

- To demonstrate compliance with environmental requirements by occupiers, planners, development agencies and developers;
- To define 'green building' by establishing a common standard of measurement and by providing a means for environmental improvement;
- To raise consumer awareness of green-building benefits;
- To provide occupant benefits to create a better place for people to work and live;
- To achieve higher rental incomes and increased building efficiency;
- To promote integrated, whole-building design practices;
- To recognise environmental leadership in the building industry and transform the building market.

BREEAM – Building Research Establishments Environmental Assessment Method[1]

BREEAM is the UK's most widely recognised measure to assess the environmental performance in environmental design and management. The method covers a range of environmental issues, presenting the results in a 'rating' that is understood by those involved in property procurement and management. BREEAM is still very much an environmental tool, predominantly focusing on energy management and carbon dioxide reduction.

BREEAM assesses the performance of buildings in the following areas:

- Management: overall management policy, commissioning site management and procedural issues;
- Energy use: operational energy and carbon dioxide issues;
- Health and well-being: indoor and external issues affecting health and well-being;
- Pollution: air-pollution and water-pollution issues;
- Transport: transport-related carbon dioxide and location-related factors;
- Land use: greenfield and brownfield sites;
- Ecology: ecological-value conservation and enhancement of the site;
- Materials: environmental implications of building materials, including life-cycle impacts;
- Water: consumption and water efficiency.

Developers and designers are encouraged to consider these issues at the earliest opportunity to maximise their chances of achieving a high BREEAM rating. Credits are awarded in each area according to performance (see Figure 7.1). A set of environmental weightings then enables the credits to be added together to produce

Category	Number of Credits	Value / Credit	Maximum Score
Management	9	1.67	15
Health & Wellbeing	15	1.00	15
Energy	17	0.83	14
Transport	13	0.83	11
Water Consumption	6	0.83	5
Materials	11	0.91	10
Land Use	2	1.50	3
Ecology	8	1.50	12
Pollution	11	1.36	15
TOTAL			**100**

Figure 7.1: How the BREEAM system works.

a single overall score. The building is then rated on a scale of PASS (25 to 40), GOOD (40 to 55), VERY GOOD (55 to 70), EXCELLENT (70 to 85) or OUT-STANDING (85 and above), and a certificate is awarded that can be used for promotional purposes.

In a similar approach to that of LEED, there is no specific scheme for refurbishments. Instead, minor and medium-level refurbishments can be assessed through the BREEAM in-use criteria detailed below, whereas the major-level changes should be determined through the BREEAM new-build categories, which include criteria for refurbishments.

BREEAM In-Use is a scheme to help building managers reduce the running costs and improve the environmental performance of existing buildings. It consists of a standard, an assessment methodology, and a third-party certification process that provides a clear and credible route map to improving sustainability. BREEAM In-Use is currently relevant to all non-domestic buildings: commercial, industrial, retail and institutional buildings.

BREEAM In-Use is designed to:

- Reduce operational costs;
- Enhance the value and marketability of property assets;
- Create a platform for landlords, owners and tenants to negotiate building improvements;
- Provide a route to compliance with environmental legislation and standards, such as energy labelling and ISO 14001;
- Ensure greater engagement with staff in implementing sustainable business practices;
- Provide opportunities to improve staff satisfaction with the working environment with the potential for significant improvement in productivity;
- Demonstrate commitment to corporate social responsibility (CSR);
- Improve organisation effectiveness.

A BREEAM In-Use assessment is broken down into three parts:

Part 1 Asset performance – the inherent performance characteristics of the building based on its built form, construction and services.

Part 2 Building-management performance – the management policies, procedures and practices related to the operation of the building; the consumption of key resources such as energy, water and other consumables, and the environmental impact, such as that of carbon and waste generation.

Part 3 Organisation effectiveness – the understanding and implementation of management policies, procedures and practices; staff engagement; and delivery of key outputs. Note: the criteria for Part 3 are based on office buildings only; other building types can be assessed but will be indicative only.

LEED: Leadership in Energy and Environmental Design[2]

The LEED Green Building Rating System was launched in 1998 and was first used to certify a project in 2000. The tool was developed by the United States Green Building Council (USGBC), which undertakes the formal assessment and certification. LEED has been adopted by state and local government for public-owned and public-funded buildings. It is used increasingly on privately owned buildings and has been adopted and used in over 40 different countries around the world. LEED is a voluntary, third-party certification process for developing high-performance, sustainable buildings. It is a nationally accepted benchmark, which recognises the design, construction and operation of green buildings.

Members of the USGBC, representing all segments of the building industry, developed LEED and continue to contribute to its evolution. LEED standards are currently available or under development for:

- New commercial construction and major renovation projects (LEED-NC);
- Existing building operations (LEED-EB);
- Commercial interiors' projects (LEED-CI);
- Core and shell projects (LEED-CS);
- Homes (LEED-H);
- Neighbourhood development (LEED-ND);
- Major renovation;
- Existing buildings;
- Schools;
- Retail (pilot);
- Healthcare (pilot).

LEED EBOM is the main rating category for small and medium-sized refurbishments, with major-scale refurbishments best assessed under the New Commercial category. In addition, LEED for Commercial Interiors (LEED-CI) can be used to certify the sustainability of tenant improvements and interior renovations.

LEED provides a complete framework for assessing building performance and meeting green building goals. Based on well-founded scientific standards, LEED emphasises strategies for sustainable site development, water savings, energy efficiency, materials' selection and indoor environmental quality. LEED recognises achievements and promotes expertise in green building through a comprehensive system offering project certification, professional accreditation, training and practical resources.

The LEED assessment tool assesses projects against credits grouped under six categories. Whilst all credits are weighted equally, the different number of credits that make up each category is, in effect, a weighting method. Whilst there are some mandatory credits shared across the LEED tools, most of the tools also have more specific mandatory criteria. The system rates buildings from 1 to 100 and is banded to give a particular rating. Each tool has slightly different scoring bands. LEED assesses the performance of buildings in the following areas (see Figure 7.2):

- Sustainable Sites – location of site (erosion, flooding, brownfield), green transport, and infrastructure;
- Water Efficiency – water-conservation measures, recycling and use of rainwater and grey-water technologies;
- Energy and Atmosphere – commissioning, renewable technologies, and process issues;
- Materials and Resources – recycling and recycled-content materials, provision of recycling space, and sourcing of materials;

Category	Possible Points (% of total)	
Sustainable Sites	14	20%
Water Efficiency	5	7%
Energy / Atmosphere	17	25%
Materials / resources	13	19%
Indoor Environmental Quality (IEQ)	15	22%
Innovation	4	6%
Accredited Professional	1	1%
TOTAL	**69**	**100**

Figure 7.2: How the LEED system works.

- Indoor Environmental Quality – ventilation, use of materials that reduce VOC emissions, and occupancy comfort;
- Innovation and Design Process – additional points for exceptional performance beyond the requirements of LEED.

The rating system is based on four levels: Certified (26 to 32 points), Silver (33 to 38), Gold (39 to 51) and Platinum (52 to 69).

Using LEED certification and registration data as an indication of green building market trends is becoming an opportunity to demonstrate environmental responsibility. Whilst existing buildings represent only 4 per cent of LEED-certified square footage, they now account for 20 per cent of LEED-registered square feet, spurred by a combination of energy prices, environmental awareness and generational shifts. LEED does not reward pre-retro-fit versus post-retro-fit improvement; it rewards a building that operates according to specific performance metrics and goals, including energy-efficiency performance, indoor environmental quality, and O&M policies and procedures that reflect environmentally responsible, high-performance building standards. In many cases, a significant energy retro-fit won't be enough to meet LEED EBOM requirements.

7.3 Project Sustainability Assessment

Each project should be reviewed for sustainability measures at the briefing stage. A generic assessment template is provided in Table 7.2, which incorporates the 14 sustainability categories present during the project activity itself, and also any changes upon completion of the project.

The assessment should include consideration at the design stage of using products and materials that will minimise the impact on the environment during their complete life cycle. There is a vast amount of information that has been documented about the choice of fabric and technology, which is constantly changing[3].

Table 7.2: Template for assessing sustainability risks during projects

Sustainability Category	Relevant (Y / N)	Design Issues	Build Issues	Operation	Performance Measures
Management					Certification Rating
Emissions to Air					CO_2 equivalent/m^2 (NIA)
Land Contamination					% brownfield development
Workforce Occupants					% occupant satisfaction
Local Environment & Community					% sustainable spend
					No. of complaints
Life cycle of building/ products					Life-cycle payback (months)
Energy Management					kWh/m^2 (NIA)
Emissions to Water					m^3/person
Use of Resources					% sustainable spend
Waste Management					kgs/spend
Marketplace					Benchmark score
Human rights					Benchmark score
Biodiversity					% managed biodiversity space
Transport					number/ workstations

At each stage of the project, the table is updated providing the high-level information, with underlying documentation, and the areas to minimise the risk. The first process is to identify whether the impact is relevant to the project – for example, a chiller-replacement project will involve hazardous materials and energy consumption, but is unlikely to have any great impact on material usage.

It is important that any dust or particulate matter that might be generated as part of a project to refurbish or construct a building is carefully controlled and managed. Any release into the atmosphere will impact on local residents and the community, and this can be acted upon by the local authority. During the design phases of the project, if the generation of dust is identified as a major impact, the local community should be included in the communication loop to gain their buy-in at this stage of the project and to make them aware of the potential for releases to occur. As with dust, each project should be risk-assessed for elements such as noise, and recommendations made to minimise potential pollution. This includes specifying working hours, the use of noisy equipment and any necessary soundproofing.

The key risks identified through the various stages should be noted down in the respective columns for Design Issues and Build Issues, to reflect the existing information available. As more information becomes available, particularly for larger projects, these risks might change. A further element that is not generally considered comprises the additional benefits that can be gained from the use of environmental best practice through the project activities. Whilst cost and times-cale are predominantly the key factors, many projects produce savings resulting from reduced waste disposal to landfill, or from improved energy efficiency and water efficiency. It is recommended to equate the benefits from the project, where performance measures can be applied to the major risks identified.

Sustainability Property Process (SPP)

What does it provide?

The Sustainable Property Process has been developed as a standardised process. As such, the process has been developed to be incorporated within an organisation's capital-approval process, though it can also operate as a stand-alone entity.

The main objectives of the SPP are to:

- Identify sustainable opportunities at key stages throughout the building's life cycle;
- Provide a standardised methodology for global implementation;
- Focus on where a difference can be made;
- Transfer sustainability knowledge throughout the facility's life cycle.

In particular, the SPP will:

- Deliver more sustainable buildings;
- Cover the various aspects of social, environmental and economic sustainability;
- Involve a life-cycle-costing approach;
- Assist in building selection and decision making.

Sustainability Measures

The SPP was designed to align with the workplace standards and therefore provide a focus on the environmental sustainability criteria. In total, eight categories were captured and used throughout:

- Management
- Energy
- Transport
- Water
- Waste
- Biodiversity
- Building/product life cycle
- Community/local environment

How does it work?

The process was developed as a cycle to transfer knowledge from the brief through the design, construction and operations' phases. Critically, a loop was also included to feed lessons learnt and performance knowledge back into the briefing stage.

A series of five tools, developed specifically, can be incorporated within the existing capital-projects process. This approach should align with the existing systems and therefore result in no additional time impacts on the project as the activities are dovetailed into the existing process. It also provides key information at each of the stages to ensure that informed decisions can be made.

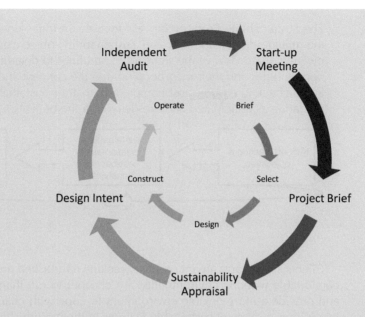

Figure 7.3: Sustainability property process.

The inclusion of sustainability criteria, from the initial kick-off meeting through to the ongoing facilities-management operations, will be set down systematically to identify and promote benefits and mitigate risks.

SPP Stage	Stage Objective	Sustainability Inclusion
Briefing	Determine project feasibility and alignment with business strategy	Determine sustainability objectives and capital-budget provisions for equipment
Concept Design	Select preferred project option(s)	Assess sustainability performance and option(s)
Design	Finalise project scope, cost and schedule and get project funding	Sustainability inclusion within project funding
Construction	Produce an operating asset consistent with scope, cost and schedule	Delivery of a sustainable building to agreed specification of design and resource efficiency during construction
Operation	Evaluate asset to ensure performance to specification and maximum return to shareholders	Post-handover performance evaluation against original criteria and ongoing monitoring and measurement

(Continued)

The Sustainability Property Process focuses on three key stages involving: the initial identification of the sustainability objectives at the kick-off meeting; the review of the short-listed buildings to determine sustainability opportunities and their incorporation into the concept and detailed design; and a check to ensure, post-completion, that the sustainability features are operating as anticipated and are delivering the benefits.

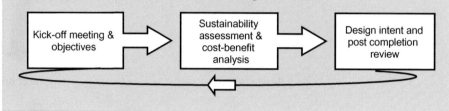

Clients and tenants are increasingly seeking refurbished properties that provide comfortable working environments, are cheaper to run than their predecessors, and provide a more flexible environment to cope with changing work patterns. In addition, a range of sustainable features in their refurbishment projects, which include the use of more 'environmentally-friendly' products in the design phase, are also being sought. There is considerable scope to use products with recycled content in the installation of partitions and building elements that are associated with dividing up the space.

Research undertaken by Davis Langdon shows that medium-sized, office-refurbishment projects offer good opportunities for product substitution. This particularly applies to refurbishments that can be classed as 'office fit-outs', involving the renewal and upgrading of internal fittings, finishes and engineering services, along with some minor structural alterations[4]. Major refurbishment works that incorporate substantial structural alterations to create an entirely new layout also offer significant opportunities, particularly in changes to the external cladding and structural elements of upper floors.

There is currently very little information available to constructors on how refurbishments can be carried out in such a way as to deliver reduced carbon emissions, (e,g. with better designed heating and ventilation equipment, or using low/zero-carbon materials, such as self-ventilating concrete blockwork). Also, builders and refurbishers should be encouraged to consider whether they can substitute primary products with secondary or recycled materials, thereby producing environmental gains for the community.

7.4 Performance Improvements from Standards

A review of scientific literature and studies indicates that green initiatives, and certifying the workplace as green, increase productivity, but the challenge is in quantifying those gains in relation to profit. The Society for Human Resource Management (SHRM) in a 2009 poll, 'Green Initiatives, What has Changed in One Year', defined a green workplace as one that is environmentally sensitive, resource-efficient and socially responsible. This definition can be viewed from the perspective of a corporate strategy, easily fitting within the concept of sustainability.

Many scientific studies have been conducted in various workplaces and educational facilities to gauge the impact on human performance, as defined by reading speed and comprehension, learning, word memory, multiplication speed, signal recognition, time to respond to signals, and typing speed. Obviously, these tasks could all contribute to productivity.

Here are some of the findings:

- There is a strong link between the quality of the indoor air and the incidence of allergy and asthma symptoms. This is significant as 20 per cent of the US population have environmental allergies and 6 per cent have asthma.
- Higher concentrations of carbon dioxide, which indicate a lower rate of ventilation, can cause fatigue, headaches and increased risk of 'sick building syndrome'. Also, tests have shown poorer performance in computerised tests of reaction time.
- Temperature matters. Performance increases with temperatures up to 70–72°F (21–22°C) and decreases with temperatures above 73–75°F (23–24°C). The highest productivity is at 71.6°F (22°C). (The optimal environment is one where the individual occupant can control the temperature.)
- Whilst studies of lighting with office workers has produced mixed results, a study performed in a school showed improvements in a standardised test of 16–26 per cent in classrooms with the most daylighting, or window area, respectively.
- Healthcare maintenance has been shown to have a strong correlation to an employee's productivity level. Successful good-health programmes, in particular, improve future productivity.

Deloitte Consulting conducted a survey of major employers in 2008 that had implemented green retro-fits[5]. Here are some of the results:

- 93 per cent of respondents reported a greater ability to attract talented employees;
- 81 per cent saw greater employee retention;
- 87 per cent experienced an improvement in workforce productivity;
- 75 per cent reported improvements in employee health;
- 100 per cent experienced an increase in goodwill/brand equity.

Several studies, including the 2003 Report to California's Sustainable Building Task Force, which involved 33 green building projects, calculate small percentage increases in productivity. The 2003 report recommends attributing a 1 per cent increase in productivity and health to LEED-certified and LEED-Silver buildings, and a 1.5 per cent gain in LEED-Gold and LEED-Platinum buildings. Those benefits resulted primarily from better ventilation, lighting and general environment. In a five-year company case study, researchers at Carnegie Mellon University measured a 3.2 per cent productivity gain, or US$1,600 per employee per year, from lighting improvements alone.

Research indicates that schemes such as LEED and BREEAM can improve workplace effectiveness and return on investment (ROI). A 2009 Michigan State study, Life Cycle Cost Analysis of Occupant Well-being and Productivity in LEED Offices, found that groups moving to LEED-certified office buildings missed less work and put in almost 39 hours more per person annually. According to the study, the total bottom-line benefits from gains resulted from fewer allergic reactions and reduced stress. The productivity boost ranged from US$69,601 to more than US$250,000 per year, the study showed.

The key to linking LEED to ROI, and to other valuable measures of organisational effectiveness, is to plan early and strategically. The experience of several leading office tenants across the country demonstrates how implementing a LEED-CI-based plan can create measurable value through green interiors. LEED-CI should be viewed as integral to project planning rather than as an 'add-on' to interiors' projects.

Amongst the benefits of LEED, cited in studies like the Michigan State report, are direct gains in overall health for office workers. These improvements come from better indoor air quality (IAQ), increased daylighting, and other changes seen to enhance morale or reduce stress. The bottom line: happier, healthier, more relaxed employees tend to produce better work.

7.5 Behavioural Change

One of the biggest barriers to sustainable refurbishment is the ability to deliver the improvement over time through behavioural change. As with any change programme, engagement of staff is critical to its ongoing success. There are significant challenges to the 'greening' of existing buildings. The reality is that green is multi-disciplinary, requiring coordination of competing priorities, from occupant comfort and productivity to energy efficiency and waste reduction. All too often, multiple stakeholders have responsibility for different green aspects of a building, whilst budgets are tight and daily workloads are increasing.

Sustainability initiatives are typically embraced by building occupants, but can be fleeting when behavioural change is involved. Whilst environmental responsibility is more visible and desired than ever, building occupants might baulk when requested to return silverware and plates to the cafeteria instead of tossing away disposables; to manually turn on lights if working later, due to lighting-schedule changes; or to give up their personal desktop printer for a networked one.

Ultimately, the greatest challenge to the 'greening' of existing buildings is financial. Whilst rising energy prices and shifting attitudes are making green efforts a greater priority, most organisations are faced with budgetary constraints and stringent return-on-investment requirements. In this context, few organisations bundle together their sustainable refurbishment activities, rather do they pursue individual measures piecemeal and justify such measures financially based on strict ROI criteria, or through a more qualitative lens such as employee morale.

When it comes to the cost of sustainability measures, some are easier to quantify than others. Energy conservation is truly the 'dollars and cents' of sustainability, and will continue to play that role in the future. In any building there is typically 'low hanging fruit' that can yield significant savings, such as lighting retro-fits, Building Automation System optimisation, or mechanical-system improvements. These are measurable savings, and can easily be translated into the carbon dioxide reduction equivalents that organisations are using as metrics today.

Savings from a retro-fit need to be assessed in conjunction with the often intangiblebenefits of sustainability – market differentiation, employee retention, customer loyalty, and worker productivity.

Endnotes

1 www.breeam.org/
2 www.usgbc.org/
3 Shah, S., *Sustainable Practice for the Facilities Manager*, Wiley-Blackwell 2006
4 Rawlinson S., Harris I. Davis Langdon, Cost Model: Office Refurbishments. *Building Magazine* 2nd October, 2009
5 The dollars and sense of green retrofits, A Joint Study by Deloitte and Charles Lockwood, 2008

Part 3

Low-Carbon Technologies and Materials

Since the energy crisis in the 1970s, building standards have consistently set tougher energy-efficiency and low-carbon criteria to be met. Building Regulations in the past twenty years have included energy targets to be met for new buildings and major refurbishments. More recently, the Stern report identified a number of measures to stabilise carbon emissions through:

- Carbon pricing through tax, trading or regulation as a foundation for climate-change policy;
- Technology policy to develop a range of low-carbon and high-efficiency technologies on an urgent timescale;
- Removal of barriers to behavioural change to encourage take-up of opportunities for energy efficiency.

These policy requirements encompass a number of areas to review the life cycle of carbon in the built asset and the means by which it can be reduced. The various Building Regulations seek to reduce energy demand at the outset of the refurbishment itself. As operational-energy use is targeted and reduced, the amount within the materials as embodied carbon increases proportionally. There is a need, therefore, to look at carbon reduction within the building holistically.

Refurbishments can lead to an increase in the energy use due to changes in the use of the building, the addition of larger numbers of equipment (e.g. computers, lighting and associated servers) or a greater number of people catered for within the building. The objective here is to reduce this additional energy use and potentially bring it below the original baseline.

Sustainable Refurbishment, First Edition. Sunil Shah.
© 2012 Sunil Shah. Published 2012 by Blackwell Publishing Ltd.

This part covers carbon reduction in buildings in four areas – energy efficiency; behavioural change; capital expenditure (renewable); and embodied carbon.

- Chapter 8 – energy-efficiency measures focused on changes to the fabric within the built asset;
- Chapter 9 – behavioural changes necessary to maintain the efficiency measures implemented;
- Chapter 10 – capital projects involving renewable-energy and low-carbon technologies;
- Chapter 11 – embodied carbon of the materials.

8 Energy-Efficiency Measures

Even though the economic environment can be difficult for the property sector, it is important to be mindful that policies and legislation relating to climate change and energy consumption will continue to advance. When the economic situation improves, property companies that have addressed the challenges of energy efficiency may find themselves better placed to take advantage of an eventual upturn in the property market.

Tackling climate change starts with energy efficiency. Reducing energy demand is the quickest and most cost-effective way of addressing carbon emissions, with existing buildings offering the biggest gains. We cannot afford to ignore the potential available from the refurbishment of existing non-domestic buildings.

Existing non-domestic buildings account for 20 per cent of the UK's total carbon emissions – more than 100 million tonnes a year[1]. More than 40 per cent of these buildings standing in 2050 will have been built before basic energy-efficiency standards were introduced in the 1970s. There is a similar story across the US, Europe and Australia, with a large proportion, in excess of 60 per cent of buildings currently built, which will still be standing in 2050.

Energy-efficiency improvements can be carried out as part of a refurbishment whilst there is vacant possession or on a minor-works basis whilst a tenant is in occupation of the building. Some improvements can be carried out during normal business hours whilst the building is occupied, however, certain improvements can only be undertaken out of hours (evenings/weekends), which would typically command a cost premium.

Owners of multi-tenant buildings or managed space, which comprise the bulk of commercial space, are primarily motivated by return on investment. To justify the costs associated with energy-efficiency retro-fits, owners must be convinced that the investment will be repaid by some combination of reduced operating expenses, higher rental rates and greater occupancy levels. The percentage of

Sustainable Refurbishment, First Edition. Sunil Shah.
© 2012 Sunil Shah. Published 2012 by Blackwell Publishing Ltd.

tenants willing to pay higher overall occupancy costs for green space is not large, and tenants that greatly value sustainability have historically gravitated towards newer buildings that have been designed and built to higher energy and environmental standards. However, following the recent introduction of financial incentives, the incremental cost of retro-fitting older buildings to achieve improved energy performance is now less than the incremental cost of achieving the same performance in a new building.

Evidence is developing from the market of a shift towards sustainable refurbishments for certain aspects. Globally, 2 per cent of buildings are built as energy-efficient green buildings and this is set to continue to increase. This chapter outlines the various factors affecting the market for sustainable refurbishments and the way in which the market is responding to these factors.

Chapter Learning Guide

This chapter will provide an overview of the energy-efficiency measures for domestic and commercial developments.

- Fabric and mechanical measures to be included capturing cost paybacks where available;
- Use of marginal-abatement cost curves and energy-performance contracting as a means to manage longer-term savings potential;
- Government support provided specifically for energy-efficiency measures aimed at cost neutrality;
- Tenant and landlord demand for energy-efficiency measures versus capital spend and obsolescence.

Key messages include:

- Energy-efficiency opportunities can provide significant cost savings and can be implemented when occupants are in situ;
- Natural ventilation and other energy-efficiency methods provide increased occupier and tenant satisfaction at reduced running costs;
- Refurbishment can lead to increases in energy use.

8.1 Introduction

The US Departments of Energy, Commerce, Labor, and four other federal agencies, banded together to establish a US$129.7-million regional research centre to develop and implement new technology for building efficiency in February 2010. With buildings accounting for almost 40 per cent of US energy consumption and carbon emissions, the effort was the newest in an array of building-efficiency initiatives that are intended to reduce energy use, emissions and property owners' utility bills whilst stimulating the economy and creating jobs. The funding will not only advance the development of new, energy-efficient technologies, it will help local governments, businesses and homeowners save money on their utility bills by putting the technology to work.

The agencies' efforts are intended to promote regional growth by establishing an Energy Regional Innovation Cluster (E-RIC), whose focal point is an Energy Innovation Hub. The regional hub, or research centre, is to develop new

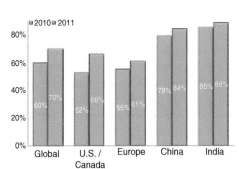

2010 Global	2011 Global	Drivers of efficiency	Europe	India	China	US/CA
①	①	Energy cost savings	①	①	①	①
④	②	Gov't/utility incentives/rebates	②	③		②
③	③	Enhanced brand or public image				③
N/A	④	Increasing energy security	③	②	②	
②	④	Greenhouse gas reduction				
⑥	⑥	Existing policy			③	

Figure 8.1: Efficiency is increasingly important (Source Johnson Controls).

technology for improved energy-efficient building systems, which are then to be put in place in structures in the area. The hub is one of three proposed by the Obama administration and funded in the budget for the 2010 fiscal year.

Interest in energy efficiency remains high, but the investments and actions to improve buildings have weakened, according to the Energy-Efficiency Indicator survey by Johnson Controls Inc. and the International Facility Management Association (see Figure 8.1)[2].

The survey has shown a decline in capital and operating expenditures for energy efficiency – a reflection of hesitancy on the part of decision-makers as a result of the current economic climate. Undertaking projects to increase energy efficiency in buildings has been the most popular strategy to achieve greenhouse-gas emissions' reductions, which had been pledged to the public. However, fewer people have said that they expect to make energy-efficiency improvements in the near future.

At the same time, many reported that they had already taken fewer steps, or none at all, toward improving efficiency in the recent past. Limited funding, a desire for greater incentives and shorter payback periods, uncertainty about energy and regulatory policy, and the changing political climate – in addition to the ailing economy – all contributed to the apparent setbacks.

Case Study

A library in Frankfurt hired an energy-management company that replaced the lighting, changed the cooling system, and modified the heating and power supply. The total cost was €600,000 (US$949,000) and resulted in a one-third reduction in the building's energy consumption and saved the library €120,000 (US$190,000) per year. The annual savings allowed the library to repay the energy-management company that financed and implemented the renovation out of its savings over ten years and, importantly, achieved the implementation without investing any capital. In addition to acquiring improved space, part of the €600,000 pays for much-needed furniture and equipment for the library and also reduces the library's energy use by one-million kilowatt-hours per year.

Buildings represent a golden opportunity for cutting greenhouse-gas emissions. 'Energy-efficiency options for new and existing buildings could considerably reduce CO_2 emissions *with* net economic benefit', according to the IPCC Fourth Assessment Report. 'By 2030, about 30 per cent of the projected GHG emissions in the building sector can be avoided with net economic benefit.' The IPCC report further remarks that it is 'often more cost-effective to invest in end-use energy-efficiency improvement than in increasing energy supply to satisfy demand for energy services' – in other words, making buildings more energy-efficient would reduce the need to build more coal-fired power plants[3].

Wherever possible, refurbishment should incorporate passive energy-efficiency measures, such as daylighting and natural ventilation, to improve energy efficiency and reduce running costs. Research has shown that, in temperate climates, despite the additional capacity and running costs, occupant satisfaction was no different from that in naturally ventilated buildings[4].

Buildings incorporating passive measures offer potential environmental benefits. The use of air-conditioning may appear to be the only choice in refurbishing buildings initially, but modifications and improvements to the fabric and insulation, and a reduction of heat gain, can dramatically reduce the need for direct air-conditioning. Typical benefits include:

- More attractive, 'daylighted' interiors;
- Less dependence on mechanical cooling systems;
- Lower energy and maintenance costs and reduced exposure to carbon taxes;
- Good long-term investment with less dependence on supplies of delivered energy;
- Less overheating, improved comfort and, possibly, a healthier internal environment;
- Opportunities for better personal control of the local environment, particularly in cellular offices.

Developers and investors have demonstrated concerns about the marketability and financial returns from passive designs, especially for premium properties. Common concerns are:

- Lower rental values; at present, passive buildings benefit from no rental premium;
- Risks to thermal comfort, particularly if occupancy and equipment levels are high;
- Unfamiliar technologies might require changes of habits from management and occupants;
- Lack of flexibility in accommodating partitioning to suit occupiers' needs (partitions might block ventilation paths and interfere with control strategies).

These concerns may be alleviated by introducing contingency plans to allow extra services to be added. However, it is likely that the regulatory pressures to reduce carbon emissions will lead to greater demand for passive designs in appropriate locations. It is important for those who operate and occupy passively ventilated buildings to understand how they work. This helps ensure that the building functions correctly in the passive mode. Otherwise, operators might assume that the measures have failed and thus allow extra services to be installed. Once this has occurred, it is unlikely that the building will revert to its passive mode of operation.

8.2 Refurbishment Options and Approach

There is an opportunity to implement a number of energy-efficient measures during building refurbishment that are unlikely to be considered during the normal operation of the building. Refurbishment provides excellent opportunities for improving energy efficiency, although it can sometimes increase energy consumption where services are enhanced, for example, by the introduction of air-conditioning. Major refurbishment will involve a significant amount of design and, potentially, a different use of the building.

Typical areas to be considered during a refurbishment will include the following, which should be used as an overriding approach to reducing the energy demand of the refurbished space.

- Consider space layout and zoning – make effective use of AC and heating installations;
- Reduce solar gain, consider:
  ○ Internal and external shading
  ○ Overhangs
  ○ Glazing orientation and scale;
- Avoid air-conditioning;
- Minimise heat gains through building-fabric improvements;
- Implement a passive cooling and ventilation strategy;
- Implement a passive solar strategy;
- Consider heat recovery within the services strategy;
- Optimise free cooling.

Table 8.1 provides a comparison between different energy-consumption areas to identify mechanisms to reduce energy use and avoid conflicts between the various energy-consuming areas.

More detailed considerations for specific areas within the refurbishment include:

- Can energy-efficient displacement ventilation be used in rooms with low heat generation, such as reception areas?
- Can passive infra-red (PIR) occupancy sensors be used to reduce office-lighting energy consumption? Occupancy, or motion, sensors are devices that turn lights and other equipment on or off in response to the presence (or absence) of people in a defined area. Some sensors also control lighting based on the amount of daylight available in their coverage area.
- Can the building be zoned with local controls in response to heating, cooling and lighting requirements? For example, multiple local thermostat controls can be used to respond to varying demand requirements. Also consider that workspace close to windows will be affected by a higher solar-heat gain, and that areas with equipment will have a higher heat gain.
- Avoid unnecessary glare issues through internal layout design. Implement glare control through blinds or internal shutters.
- Furniture and partitions need to be positioned so as not to restrict air flow.
- Consider the level of lighting luminance that is required or preferred. If only very few individuals need extra light, can this be provided more effectively on an ad-hoc basis? Alternatively, consider dimming by lighting control linked to photocell sensors.

Table 8.1: Energy-efficiency assessment matrix (adapted from CIBSE Energy-efficiency Guide)

	Cooling	Heating	Electric lighting	Daylight / glazing	Natural ventilation
Heating	Avoid simultaneous heating and cooling				
Electric lighting	Reduce incidental gains from lights to minimise cooling	Include contribution of lighting towards heating			
Daylight / glazing	Minimise solar gains to reduce cooling loads	Minimise heat loss and maximise useful heat gain through glazing	Use suitable switching and daylight linking controls to minimise use of electric lighting		
Natural ventilation	Consider mixed-mode to use natural ventilation and avoid mechanical cooling where possible	Account for effect of open windows		Balance solar gains from glazing with increased natural ventilation. Avoid conflicts between	
Ventilation and air	Use free cooling and 'cool' recovery	Use heat recovery	Reduce electric lighting to reduce loads on air conditioning	Solar gains from glazing may increase loads on air conditioning. Heat loss may require simultaneous perimeter heating and cooling	Use natural ventilation instead of air conditioning where possible, or consider mixed-mode ventilation

- Access to opening windows or vents enhances occupant comfort by giving them control of the fresh-air intake.
- Has a mixed-mode strategy been considered? This energy-efficient system will allow an automatic control system to combine mechanical ventilation with natural ventilation.

- Identify the expertise that the occupier is likely to require to operate, maintain and control mechanical and electrical systems and other moving parts, so that 'design for manageability' can be realistically undertaken. This will also influence the degree of user training required for operation of the BMS and controls.

Pulling these various strands together helps to inform the energy-efficiency options that are available at the four levels of refurbishment. There will be a degree of interchange between the options dependent on the specific criteria of the project and on which measures prove to be most cost-effective. A template in Table 8.2 is provided to capture the fabric and mechanical measures that can be made. However, this provides only a starting point for what can be delivered and achieved based on the scale of the refurbishment.

Levels 1 and 2: Light Touch/Minor Refurbishment

Minor refurbishment generally involves refitting the interior and making minor alterations to space layout and plant, and it can present opportunities for introducing specific energy-saving measures including:

- Changing space layout to enhance daylight, ventilation and zone controls;
- Improving lighting arrangements, including automatic controls and energy-efficient lamps (lighting is often the largest single end use of energy in office buildings);
- Improving window performance by adding blinds and so on;
- Using lighter-coloured interior surfaces and furnishings to enhance the lighting efficiency.

Level 3: Extensive Refurbishment

Extensive refurbishment often involves replacement of major plant and can include some changes to the fabric, for example, window replacement. It often allows significant changes to building services' strategies. Energy-saving opportunities within extensive refurbishment include:

- Adding additional insulation;
- Adding atria and sun spaces to increase natural ventilation and daylight;
- Increasing the use of passive measures or mixed-mode strategies in air-conditioned buildings;
- Maximising use of free cooling;
- Removing (fully or partially) air-conditioning through changes to fabric, lighting and controls, for example, in shallow-plan buildings on relatively quiet sites;
- Selecting efficient plant and flexible controls, including zone controls;
- Specifying an efficient and fully insulated hot-water system, with consideration given to localised water heating where this will help to reduce standing losses.

Level 4: Comprehensive Refurbishment

Comprehensive refurbishment generally involves total replacement of plant and major changes to the building fabric, possibly only retaining the facade or

Table 8.2: Summary of energy-efficiency measures available in refurbishments

FORM AND FABRIC	
Insulation	Insulation levels to maximise thermal capacity
Windows and Glazing	Low-level horizontal pivot/sliding windows Double/multiple glazing Window frames with thermal brake Coating on glass to reduce radiation heat transfer Insulating layer (i.e. argon) in sealed glazing
Thermal Provision	Building mass to temper rate of heating/cooling
Fabric	Inclusion of low U-value materials throughout Avoid excessive thermal bridging Reduce air-permeability levels Door closers Draught lobbies
SERVICES	
Daylight	Good distribution of daylight through: • Splayed reveals • Light shelves • Prisms Introduction of daylight into deep-plan rooms by means of: • Lightwells • Atria Avoid dark internal surfaces Daylight sensors + dimmers to zone electric lighting in relation to availability of natural light and to take account of: • Daylight availability • Workstation layout • Provision of manual & automatic controls Active shading to avoid solar overheating
Management	Building Energy Management Systems Zoning (lighting and heating/ventilation)
Lighting	Lighting to achieve best practice lux levels to avoid excessive luminaire heat gains Timers and sensors
Heating and Distribution Systems	Minimisation of number, and location, of plantrooms to reduce potential losses Reliable and accurate space-heating controls Insulation of distribution systems Heat recovery for space heating and pre-heating of domestic hot water
Ventilation	Natural ventilation strategies including: • Stack-assisted ventilation (possibly via rooflights) • Trickle vents • Windows with high usability • Night-time purge options • Chilled beams Mixed-mode ventilation VRF

structural frame. It nearly always involves radical strategic changes to the building services, which provides major energy-saving opportunities, in addition to the above, including:

- Introducing passive measures to reduce external heat gains whilst maximising daylight, for example, replacing windows, providing shading, introducing atria and rooflights;
- Changing the ventilation strategy to minimise the use of mechanical ventilation;
- Assessing the need for air-conditioning, leading to a reduction and, sometimes, complete avoidance;
- Upgrading fabric thermal performance, airtightness and heating controls to reduce heating-energy requirements;
- Installing energy-efficient plant, such as condensing boilers and combined heat and power (CHP);
- Installing energy-efficient lighting and lighting-control systems;
- Improving building services' monitoring and controls, possibly through the introduction of a building-management system (BMS).

The American Institute of Architects developed a list of fifty measures[5] to reduce heat load and energy demand within a building (see Figure 8.2). The measures

1. Active solar thermal systems	26. Integrated project delivery
2. Alternative energy	27. Life cycle assessment
3. Alternative transportation	28. Mass absorption
4. Appropriate size and growth	29. Material energy and embodied energy
5. Building form	30. Natural ventilation
6. Building monitoring	31. Open, active daylit spaces
7. Building orientation	32. Passive solar collection
8. Carbon offsets	33. Photovoltaics
9. Cavity walls for insulating airspace	34. Preservation/reuse of existing facilities
10. Cogeneration	35. Radiant heating and cooling
11. Conserving systems and equipment	36. Renewable energy resources
12. Construction waste management	37. Rightsizing equipment
13. Cool roofs	38. Smart controls
14. Deconstruction and salvage materials	39. Space zoning
15. Daylighting	40. Staff training
16. Earth sheltering	41. Sun shading
17. Efficient artifical lighting	42. Systems commissioning
18. Efficient site lighting systems	43. Systems tune up
19. Energy modeling	44. Thermal bridging
20. Energy source ramifications	45. Total building commissioning
21. Energy-efficient appliances and equipment	46. Vegetation for sun control
22. Environmental education	47. Walkable communities
23. Geoexchange	48. Waste-heat recovery
24. Green roofs	49. Water conservation
25. High efficiency equipment	50. Windows and openings

Figure 8.2: American Institute of Architects 50 to 50.

captured not only those factors described above, but also external factors, including vegetation to increase summer shading, and other areas to be discussed later including renewable energy and education.

8.3 Assessing Costs of Energy-Efficiency Measures

The cost-effectiveness of energy-efficiency investments can be expressed in several ways. The simple payback period is the initial cost divided by the monthly, or annual, cost savings. Discounted cash-flow methods are usually used in larger organisations and for major investments, although life-cycle costing is becoming more common.

The capital cost of energy-saving initiatives taken during refurbishment is often reduced because they are easier to install as part of refurbishment work rather than when the building is operational. In addition, the costs of disruption and interference with normal working practices are minimised. For measures included during refurbishment, where plant is to be replaced, the true cost of installing energy-efficient plant is only the over-cost compared to the minimum required for regulatory compliance, for example, the extra cost of a high-efficiency 'condensing' boiler compared with that of a conventional boiler. This approach substantially reduces payback periods and increases the cost-effectiveness of energy-efficiency measures. Therefore, an option to meet regulatory compliance as a minimum over a period of time should also be costed to provide a baseline from which any of the low-carbon options can be assessed.

In some cases, energy-efficiency measures can reduce overall capital costs, for example, the provision of natural ventilation that avoids the need for air-conditioning. Some energy-efficiency measures may qualify under the government-funded incentive schemes designed to encourage the installation of energy-efficient equipment. Measures can also be divided into those that require limited attention, such as insulation, and those that require active management to ensure that they continue to be effective, such as building-energy-management systems (BMS). For the latter, manageability is as important as the potential for energy savings, and should be carefully considered in the design.

The IPF have researched the capital cost of making energy-efficiency improvements beyond current market standards[6]. This involved estimating the extra capital cost of the improvement compared with making a like-for-like replacement to current market standards. For example, an 85-per-cent-efficiency gas-fired boiler would be the current market replacement for a defunct early 1990s boiler. Going 'beyond current market standards' would entail installing a 90-per-cent or 95-per-cent-efficiency condensing boiler.

Maintenance costs, and the cost of the energy saved from each efficiency improvement, were estimated for the purposes of calculating the internal rate of return (IRR) and the time taken to pay back the initial investment. A notional discount rate of 7 per cent was used to discount the cash flows for each improvement. These payback periods showed little sensitivity to changes in the discount rate applied.

The results of an IPF[7] study showed that there is a broad consistency across all buildings in terms of the energy-efficiency technologies that are most cost-effective relative to the amount of carbon dioxide saved. Many of these technologies could also form part of a replacement programme whilst the building is in occupation instead of being delayed until the next major refurbishment. This will be important where a planned refurbishment is some years away and where one or more leases are due to expire. All the improvements are tried-and-tested solutions for reducing energy consumption and carbon dioxide emissions from buildings. Incorporating these technologies into a refurbishment, and therefore incurring an increased capital cost, achieves the reduction in carbon dioxide emissions outlined previously. The energy saved from making many of these improvements provides significant IRRs to assist the business case. However, the traditional landlord–tenant relationship means that the landlord will not directly benefit from making the energy-efficiency improvements, which therefore reduces the attractiveness to a potential investor.

The research demonstrates that there are a number of quick gains, arising from the more strategic refurbishment decisions that can be taken by all stakeholders within the property and investment communities to reduce the risk of obsolescence of existing buildings caused by the tightening of legislation, thus increasing tenant demand. Having better data on the costs and efficiency savings of these improvements is also expected to be instrumental in enabling landlords and tenants to more effectively negotiate terms and conditions to support more sustainable property occupation and management.

The carbon dioxide emissions of all existing office buildings can be vastly improved through refurbishment. The savings are more significant for older offices compared to recently constructed buildings. Table 8.3 summarises the reduction in baseline carbon dioxide emissions by improving existing offices to current market standards and subsequently spending extra amounts.

The key findings were:

- Modernising older offices to current market standards reduces baseline carbon dioxide emissions by approximately 25 per cent.
- Refurbishing all older offices to modern standards reduces baseline emissions by approximately one-quarter and can be achieved with no additional expenditure above the cost of a market-standard refurbishment. This upgrade is estimated to be sufficient to achieve an Energy Performance Certificate rating of C.
- Additional expenditure of £50/m² reduces total baseline emissions by approximately 50 per cent for older offices.
- Additional expenditure of £50/m² when undertaking a refurbishment of 1990s-offices achieves a baseline reduction of between 42 per cent and 51 per cent. It is estimated that this additional investment could be sufficient to achieve an EPC rating of B for buildings that are starting from a baseline rating of D, or

Table 8.3: Cumulative percentage saving of CO_2 against baseline for a range of office buildings[8]

	1990 Office	2005 Office	Pre-1940 Office	EPC Rating
Market level	25 per cent	0 per cent	30 per cent	C
£25/m² budget	37 per cent	14 per cent	47 per cent	B/C
£50/m² budget	46 per cent	28 per cent	51 per cent	B
£75/m² budget	49 per cent	29 per cent	55 per cent	B
£150/m² budget	54 per cent	36 per cent	63 per cent	B

Table 8.4: Cumulative percentage saving of CO_2 against baseline for supermarkets/ warehousing

	Supermarket	Industrial/Warehouse
Market improvement	12 per cent	35 per cent
£10/m² budget	23 per cent	41 per cent
£100/m² budget	40 per cent	62 per cent
£130/m² budget	47 per cent	66 per cent

even E. Just for an additional spend of £25/m² the total carbon dioxide saved from baseline emissions will range from 35 per cent to 47 per cent. The saving is less for the modern office at all levels of additional expenditure because the improvements are being applied to a building that is already more efficient.

- Based on the results of this study, older office buildings present the best opportunity to reduce carbon dioxide emissions through a refurbishment. Significant savings can be achieved for the 1990s-offices and older period offices, either for air-conditioned or for non-air-conditioned buildings. In particular, offices that are non-air-conditioned and have poor external fabric in thermal-performance terms produced the biggest reduction in emissions compared to the baseline. It was found that buildings constructed more recently had less scope to reduce baseline carbon dioxide emissions.

The energy efficiency of both types of building can be improved significantly by upgrading the buildings to market standards, particularly by spending an additional £10/m² above the cost of typical refurbishment. The baseline emissions for the industrial/warehouse building are half those of the supermarket. The latter should therefore be targeted to reduce their emissions because the quantity of carbon dioxide saved can be significant. The substantial improvement in baseline emissions for industrial/warehouse units, arising from spending only a small additional amount, means that these buildings also present a key opportunity.

Installing variable-speed pumps is a key opportunity together with improving the efficiency of the lighting. Upgrading air-conditioning pumps to a variable-speed type ranked as the third-most cost-effective improvement for buildings where air-conditioning energy consumption is very significant. This included the post-2002 office and the supermarket. Variable-speed air-conditioning pumps are less important for older office buildings where heating demand is a bigger concern.

Marginal-Abatement Cost Curves

There are a number of analytical, high-level, energy-saving tools available from McKinsey & Company, RMI, and from Lawrence Berkeley National Laboratory, all of which find sustainable refurbishments to be an area where significant cost savings can be made. An increasingly well-recognised tool is the marginal-abatement cost curve (MACC), which captures the cost associated with a variety of activities to reduce greenhouse gases. A typical MACC graph is provided in Figure 8.3, which is focused towards the global buildings' sector.

MACCs summarise the technical opportunities that exist to reduce greenhouse-gas emissions based on a specific cost per tonne of avoided emissions – in the

Global GHG abatement cost curve for the Buildings sector

Societal perspective; 2030

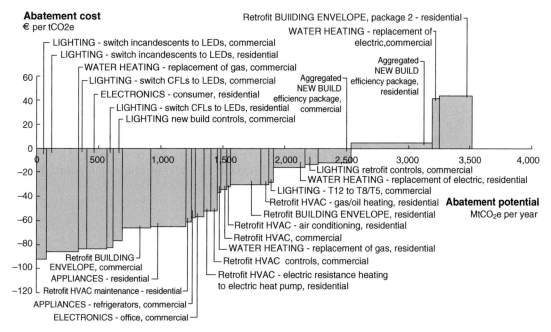

Note: The curve presents an estimate of the maximum potential of all technical GHG abatement measures below €60 per tCO₂e if each lever was pursued aggressively. It is not a forecast of what role different abatement measures and technologies will play.

Figure 8.3: Greenhouse gas abatement curve for the buildings sector. Source: Pathways to a Low Carbon Economy, Version 2 of the GHG abatement cost curve, McKinsey & Company, January 2009.

case of the example, a figure of €60 per tonne has been applied. The width of each bar represents the potential of the opportunity to reduce greenhouse-gas emissions in a specific year, compared to a business-as-usual approach. The height of each bar represents the average cost to reduce greenhouse-gas emissions by one tonne by 2030 in terms of 2005 cost. The bars on the left identify the greatest opportunity, depicting significant savings, moving to the bars on the right, which are the more expensive and less viable solutions.

There are a number of issues affecting a global approach to the cost curve and, therefore, for a country-by-country review; the curve should be read as a comparison between one technology and another as opposed to absolute-reduction figures. The curves, by being based on a longer-term, cost-saving potential, might have limited use in multi-tenanted buildings where the payback timeframes are relatively short.

Many of the cost-saving measures are closely linked to the refurbishment of buildings and, therefore, some of these measures can be directly implemented. Key areas for potential reduction include:

- **Retro-fit Building Envelope** is focused on improving the airtightness of buildings, which results in the knock-on benefits of reduced heating and cooling, the requirement for smaller plant, and fewer variations to the building, leading to improved occupant satisfaction;

- **HVAC Replacement** with high-efficiency systems, when appropriate, together with adjustments tailored to building occupancy;
- **Water Heating Replacement** with high-efficiency systems with tank-free or condensing heaters; utilising renewable technologies including solar water heating and heat pumps can further improve savings;
- **Lighting** replaced with energy-efficient light-emitting diodes (LEDs), or T5s where dimmable or photo-sensors are used.

The approach globally will differ with developed countries requiring a significant level of refurbishment based on the criteria above, and developing countries with a higher proportion of new build seeking to integrate these items into the construction process.

8.4 Fabric and Passive Energy-Efficiency Measures

Fabric improvements include the replacement of existing glazing with either double-glazed or triple-glazed windows, and retro-fitting wall insulation to the internal face of external walls. Implementing glazing improvements saves substantial amounts of carbon dioxide and replacing 'leaky' single-glazing with double-glazing improves the internal comfort of occupants by reducing draughts. Significantly, it is usually cheaper and more practical to firstly minimise carbon dioxide emissions through passive and efficiency measures before renewable-energy systems are applied. Replacing glazing is one way to save energy by reducing heat loss.

However, the cost of these improvements is considerable. Retro-fitting wall insulation within the air-conditioned offices produces a minimal saving without further modifications to the HVAC because the reduction in heat loss is countered by the requirement for increased cooling. However, for an office with only heating and no air-conditioning (such as a period office), the addition of insulation has a greater impact. The disadvantage with retro-fitting insulation in an office is the potential loss of usable floor space, which would not be viewed positively by an investor.

Insulation materials such as cellulose, made from recycled paper coated with a non-toxic fire or mould compound, or batt insulation made from recycled cotton, are sustainable materials and are thermally efficient.

Windows

Windows are one of the most energy-transmitting elements of a building, meaning that a significant amount of energy and money could be escaping from poorly performing windows. Historically, windows have been relatively small, limited to reduce heat gain into the building and also heat loss from the building, whilst also allowing daylight, air and views out. With the advent of control systems together with hermetically sealed buildings conditioned by central air systems, windows, and the use of glass, became more significant. These buildings often had a negative impact on the occupant and led to 'sick building syndrome'. Many

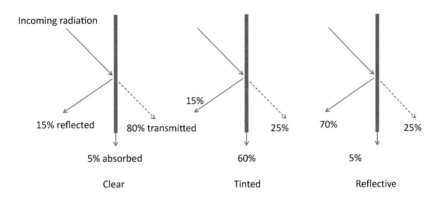

Figure 8.4: The transmission of energy through different types of glass.

of these highly glazed buildings built from the 1950s now require refurbishment. Fortunately, the appreciation of the benefits of windows has come full circle, with their value in providing daylight and air, together with providing greater energy efficiency, being better understood. An example of such an installation is the replacement of the windows within the Empire State Building.

The challenge is therefore to manage these two conflicting aspects regarding windows – to make them large enough to bring in sufficient natural daylight but also small enough to manage the heat gain and loss within the building. Figure 8.4 describes the transference of energy, from light, through three different types of window. There are a number of routes that the light energy can take – some is reflected back, some is absorbed by the window, and some is transmitted through the window and into the building. The figure shows that commonly used clear glass has almost 80 per cent of the light energy transmitted into the building, leading to significant heat gain. When either tinted or reflective glass is used, this drops dramatically to 25 per cent. However, 60 per cent of the energy in the tinted glass is absorbed, and this causes the heating up of the actual glass and frame, which also heats up the building[9]. Modern, high-performance glass enables the provision of daylight with the reflection of heat from the glass reducing heat gain and allowing larger-sized windows than would otherwise be feasible. However, care should be taken with some inert-gas-filled windows where the gas can leach out after around five years.

Particular issues arise from single-glazed, steel casement windows, which are one of the worst performers. In this case, there are few options but to replace the windows where possible. Similarly, secondary glazing should be installed to such windows if they form part of curtain walling; for the hard-to-treat or hard-to-access elevations, shutters should be provided.

The use of shading is now widely recognised, and its presence in many older buildings demonstrates its effectiveness in reducing unwanted heat gain at certain times of the day and year. Shading devices can also provide daylight redistribution, particularly in side-lit rooms where lighting is generally poor. In this situation, louvres can redistribute the light into the room, enabling a reduction in artificial light combined with increased occupant performance.

Where shading may already have been provided as an addition to the building, it must be ascertained whether the installation is integrated with the window design or whether it causes any obstruction to, or interference with, the heating and cooling systems. There are four approaches to shading:

1. External – includes overhangs, louvres, fins and blinds. Overhangs and some louvres provide effective shading together with views out, which many of the other types do not.
2. Internal – traditionally retro-fitted venetian blinds or roller blinds. These can be used to improve daylighting by redirecting light upwards and also to reduce heat gain in the building. They are dependent on human control and therefore can be left closed whilst artificial lighting is used as a proxy.
3. Inter-pane – provision of shading devices located within the middle of the window, typically in a double-skinned scenario. Criteria are similar to external devices.
4. Integrated – cover s light-shelves and prismatic systems.

Atria and Double Skins

The provision of atria and double skins does little to deliver energy-efficiency improvements to the majority of buildings. Typically, in the case of atria, the space requires to be heated and cooled in the same way as the rest of the building albeit to a lesser degree. Automatic venting of warm air through the roof is common but heat loss and gain to the building itself are equally common.

Atria can also be used to promote ventilation through the stack-effect, encouraging warmer air to be pulled through and out of the atrium space. Mechanical ventilation is necessary to achieve this, together with some form of control and monitoring to achieve best results. Since the atrium will act as a space between the daylight and the workspace, there will be a marked reduction in the daylight received within the building. Double skins provide much the same approach and performance and can therefore be treated as a certain type of atrium.

Airtightness and Ventilation

The airtightness of the building determines how much air leaks into, and out of, the building. In very leaky buildings, heat losses due to air infiltration can account for up to a third of the total heat loss, leading to their being costly to run and uncomfortable in which to work or live. Ventilation is desirable and necessary to remove moisture and other pollutants from the air. Ventilation should be provided by controllable openings, such as trickle vents or mechanical extract fans.

Some simple measures can be taken to improve the airtightness characteristics. Draught-proofing is one of the most inexpensive, yet effective, ways of making efficient use of energy in all types of building. It will often pay for itself within a year. Draught-proofing around windows, access hatches and external doors, and sealing around service pipes that pass thorough floors, walls and roofs are typically standard. Adequate ventilation is as important as draught-proofing; it should be ensured that there are working extractors in toilets, kitchens and plantrooms.

Controlled ventilation is necessary to provide a healthy internal environment, but it should be provided in an energy-efficient way. Ventilation can be provided by mechanical fans, or by passive systems that rely on the natural buoyancy of warm air to remove moisture. A ventilation strategy should include a means to provide background ventilation, for example, trickle vents in windows (a small slot in the window frame with a closable cover that allows a flow of fresh air to the room); a means to rapidly ventilate a room, such as an openable window to quickly remove odours or air from a room; and, finally, a means to extract warm, moist air from kitchens and showers to reduce the risk of condensation forming.

External Walls

Heat is lost through external walls; wall insulation has a number of benefits as detailed below. In many places, all three benefits may occur during the period of a single year.

- Heat retention in cool conditions;
- Heat exclusion during warmer conditions;
- Prevention of solar gain by the transfer of heat from a warming external wall.

There are three main methods of insulating the wall. The single most cost-effective measure is to install cavity-wall insulation. Alternatively, if the wall does not have a cavity, or if it is not suitable for filling, then either an external wall-insulation system can be applied, or an internal insulated dry-lining can be installed.

Cavity-Wall Insulation The majority of masonry cavity walls built since the 1930s are suitable for filling with cavity-wall insulation. It is cost-effective to install and can provide up to 85 per cent improvement in wall performance, however, it is not common in commercial applications. Insulation, either polystyrene beads or mineral fibre, is blown into the wall cavity through a number of holes drilled in the exterior of the wall. Once the insulation is installed, these holes will be filled up with mortar. Before the insulation is installed, a survey should be carried out to confirm that the wall is suitable for insulating and, if any defects are identified, these should be corrected first.

There is a potential for severe issues if cavity-wall insulation is installed within timber-framed structures, and also issues with as-built cavity fill, mortar snots, wall ties, and other debris. There can be an increase in cold-bridging as a result of the installation of cavity-wall insulation and also an increase in condensation and transfer of moisture as the initial moisture-isolating basis of the cavity is compromised.

The refurbishment of composite and lightweight walls may involve the stripping of either the inner or outer leaf, providing the opportunity to fix semi-rigid insulation whilst still maintaining a cavity.

Internal Insulation Where the walls are solid masonry or where a cavity wall is not suitable to receive cavity-wall insulation, then it can be cost-effective to insulate the internal surface of the wall. This is most often done by applying an insulated dry-lining, that is, a layer of plasterboard with a rigid-insulation material bonded to it. This is most cost-effective when the existing plaster is being renewed, or when existing services or fittings, such as kitchen units, are being replaced. Applying the insulation internally has the benefit of ensuring that the wall surface warms up quickly, thus helping to improve thermal comfort. It has the disadvantage of reducing the size of the room by the thickness of the insulation. The insulation layer can be up to 50mm thick. This also requires all electric sockets, switches and pipework to be relocated, once the insulation work is complete. There may also be issues with interstitial condensation, forming in the structure, which needs to be addressed at the design stage. The condensation is formed by water vapour condensing in the structure and can be avoided by applying a vapour-check barrier to the warm side of the insulation.

External Insulation   An alternative to installing insulation on the internal wall surface is to install it externally. The design and installation of external wall insulation is a specialist job, and is generally the most expensive of the three methods to insulate a wall; however, it can be cost-effective where replacement of the render is being carried out or where a major refurbishment is planned. External wall insulation consists of a layer of insulation (either rigid or mineral fibre) fixed to the external surface of the wall with a variety of decorative finishes applied to the insulation.

The external insulation can reduce the solar gain from the external wall, and, importantly, can help to eliminate and avoid cold bridges where significant heat loss can take place. Additional benefits are that the works can proceed during operational hours, minimising disruption to the building, and that the insulation enables the provision of a decorative external envelope, giving lasting weather protection.

The most common type of external insulation is insulated render but various panel systems are also available. It is important to ensure that ventilation apertures are maintained since the insulation allows the fabric of the building to act as a heat store, increasing thermal efficiency, condensation and mould growth. From a design perspective, it is necessary to consider the detailing of the eaves and to ensure that movement joints are provided on large elevations.

Floors

The amount of heat loss through the ground floor will depend on the size and shape of the floor. Generally, it is not cost-effective to insulate a ground floor unless other works are being carried out, such as the complete replacement of floorboards in a timber, suspended floor. Insulation to a timber floor can be placed within the floor thickness and supported on battens or netting, or placed above the floor when the level of the floor is being raised. If the floor is a solid concrete floor, then the only option is to place the insulation above the floor. The insulation used must be rigid, to be able to support the loads put onto the floor. The insulation can be finished with either a concrete screed or a timber-board finish. Raising the floor level may be undesirable if it disturbs the original features, such as a stone floor, or involves altering existing doors.

Where there is a floor over an unheated space, such as a garage, heat losses can be reduced by insulating the floor. Where there is access from below, this can be cost-effective. Mineral-fibre insulation can be placed between the floor joists and supported on netting fixed to the underside of the joists.

Typically, insulation is provided within intermediate floors or suspended, ground floors, fitted between the suspended-floor joists or under raised, access floors. It is important that the ventilated airspace is maintained and the need for a vapour barrier should be specifically assessed. In instances where underfloor heating is chosen, adequate crawl-space will be necessary. Similarly, access to any plumbing will need to be provided and it is also necessary to consider how any water leaks will affect the insulation.

Where the floor is raised or finished in timber above the insulation, whereby there is no connection between the floor surface and the structure, there will be significant implications for the thermal performance of the building. Since the floors will respond more quickly to changes in temperature, it will mean that additional solar gain will need to be responded to quickly, through cooling programmes.

Roofs

Refurbishing roofs can arise from the need to improve the weathering element, to improve the thermal performance or to repair the roof itself. Improving the thermal performance of the roof will mainly benefit the floor below, which will be a factor in multi-tenanted buildings.

In order to reduce the amount of heat loss through the roof, it should be insulated. Generally, roofs can be pitched, with an accessible loft space; they can have a void (space); or they can be a flat roof with no loft space.

Pitched Roofs: Insulation is generally placed between the ceiling joists to cover all of the ceiling area. All pipes in the loft space should be insulated and electric cables routed above the insulation to prevent their overheating. The insulation should be at least 250mm thick and preferably laid in two layers, one between the ceiling joists, and one across the top. This helps to further reduce the heat loss through the roof structure. To ensure that the loft space stays free of condensation, a free flow of air is required above the insulation. To maintain this flow of air, special trays should be installed at the point where the walls and roof meet, which will allow air to pass over the insulation. Finally, don't forget to insulate the loft hatch and draught-proof it as well.

Flat Roofs: Insulation in a flat roof may be within the roof structure, often called a 'cold deck', or placed above the structure as part of the roof finish, referred to as a 'warm deck'. It is most cost-effective to provide, or replace, the insulation in a flat roof when the roof covering needs renewing. With a cold-deck construction, it is important to maintain a flow of air above the insulation to remove any moisture. Owing to the problems associated with doing this satisfactorily, it is often simpler to provide insulation above the roof structure and make the roof a warm deck. The type of insulation used should be a rigid-board material, such as extruded polystyrene, and it should be at least 100mm thick.

However, the design of the roof may limit the total thickness of insulation that can be installed. If a flat roof is being converted to a pitched roof, then the roof should be insulated to the same standard as that for a new building.

Where possible, the colour of the roof should be lightened, which will reflect the sun's energy rather than absorb it, which darker materials do[10]. The higher temperature associated with darker roofs can break down roofing materials and shorten their life. So-called 'cool roofs' can also reduce the heat-island effect, which, in turn, leads to a greater level of summer cooling. Other types of roof covering, including green roofs, are discussed in Chapter 14.

Passivhaus for Domestic Refurbishment

http://www.passiv.de/01_dph/Bestand/EnerPHit/EnerPHit_Criteria_Residential_EN.pdf

If the criteria for Passive Houses are met by an energy-relevant modernisation, then an old building can also be certified as a 'Quality-Approved Passive House', based on the same criteria as those for new buildings. However, for older buildings, it is often difficult to achieve the Passive House standard with reasonable effort. The use of Passive House technology for each building component in such buildings does lead to considerable improvement in respect of comfort, structural protection, cost-effectiveness and energy requirements. The current certification criteria (to be found at www.passiv.de) are applicable, using the calculation method described in the Passive House Planning Package (PHPP) handbook and the PHPP programme.

1 Energy balance
The energy balance of the modernised building must be verified using the latest version of the PHPP. For the specific space-heating demand, the monthly, as well as the annual, method can be applied. If the ratio of free heat to heat losses is more than 0.70 in the annual method, then the monthly method should be used. The reference value (treated floor area or TFA) is the net living area within the building's thermal envelope based on the living-space regulations in Germany (WoFIV). The whole of the enclosing building envelope, for example, a row of terraced houses or a multi-storey bulding, can be considered for calculating the specific values. An overall calculation or weighted average values of several partial zones can be used to verify this. Combining thermally-separated buildings together is not permissible. Buildings that adjoin other buildings (for example, in city housing) must have at least one external wall, one roof surface, and a floor slab or basement ceiling, in order for them to be certified individually.

2 Restriction to existing buildings
Only such buildings will be certified (EnerPHit certification) for which the continued use of existing building elements would pose such substantial problems, for the energy-relevant modernisation, that modernising to Passive House level would not be practicable or cost-effective.

3 Heating demand
Alternatively, certification can be issued if the criteria for individual building components are met. In this case, the requirement for the heating demand does not apply.

4 Primary energy demand
The requirements apply to the total sum of the systems for heating, hot water, cooling, auxiliary and household electricity.

5 Summertime comfort

Excessive-temperature frequency (>25 °C) ≤10 per cent: if calculating the excessive-temperature frequency is not possible due to very high, daily temperature fluctuations, a warning appears in the PHPP 'Summer' sheet.

6 Moisture protection

All standard sections and connection details of the building must be planned and implemented so that there is no excessive moisture on the interior surfaces or in the building-component build-ups. The water activity of the interior surfaces must be kept at aw ≤80 per cent.

7 Airtightness

The airtightness of the building must be verified using a pressurisation test based on DIN EN 13829. If the value of 0.6 h-1 is exceeded, a comprehensive search for leakages must be carried out within the framework of the pressurisation test, and each relevant leak, which can cause building damage and affect thermal comfort, should be rectified.

8 Windows

It is strongly recommended that window frames that have been certified as 'Passive-House-suitable components' and triple, low-e glazing (or equivalent) are installed – using the installation principles recommended by the Passive House Institute (PHI). If this recommendation is not complied with, evidence of the comfort, according to the conditions in DIN EN ISO 7730, should be provided, or the low temperatures occurring near the window areas should be compensated by the use of heaters.

8.5 Energy-Efficiency Measures – Mechanical

Sustainable buildings utilise passive strategies to reduce energy demands where possible by maximising free cooling, natural ventilation and lighting, and natural heating due to orientation. Advanced systems are able to build upon and develop the passive measures to deliver a comfortable and effective building. The ability to retro-fit existing buildings with passive heating and cooling measures is not straightforward and can lead to increases in energy use through conflicting systems.

For cooling, measures such as openable windows, exposure of high thermal mass (uncovering concrete), night-time flushing and evaporative cooling systems will all help to stimulate and maintain passive cooling. These measures will help to generate the critical convection currents, from the exterior through the interior to an exhaust, cooling occupants along the way.

Passive-heating measures utilise the sun's energy to heat the internal space. Many buildings can adopt such an approach, but where buildings have a high internal heat load, from people or equipment, there can quickly be an overloading of heating. Knowledge about the use of the building is therefore critical to determine whether passive heating will be a benefit.

Variable-Speed Pumps and Motors

Replacing fixed-speed heating pumps with variable-speed pumps is the most cost-effective measure that can be implemented as part of a refurbishment project and it can also take place when a building is occupied. The replacement of the pumps can be undertaken during normal working hours during summer when the heating system is not being used, thus minimising disruption to the occupants. The electricity saved by using variable-speed pumps will pay for the additional capital cost after the first year and will give an IRR of more than 100 per cent.

A high-efficiency replacement motor should be sought to optimise the efficiency of the fan or pump within the system design. Where possible, this should be direct-drive rather than belt-drive.

Lighting

A significant proportion of energy bills, particularly in offices and educational facilities, comes from the cost of lighting. Older-style lighting, including halogen bulbs, are recognised to be inefficient, with up to 90 per cent of lighting power lost through heat, rather than through the provision of light.

The benefits of lighting retro-fits include improved energy efficiency and cost savings which, when combined with any available utility rebates, will minimise the payback time for the project. Above and beyond that, benefits will continue to be enjoyed from reduced energy bills, a longer life for lamps and ballasts, and improved illumination. In addition, the decrease in wattage from the improved lighting decreases the HVAC cooling load for a given space. Every watt of lighting reduction is equal to a cooling reduction of 0.33 watts, thus decreasing cooling costs. Replacing high-wattage, filament lamps or tungsten, halogen lamps with compact, fluorescent lamps or with metal, halide lamps can achieve energy savings of between 65 and 75 per cent (see Figure 8.5).

The first priority for lighting is to maximise the available daylighting through windows. Where the refurbishment allows, an increase in the available window space, or in the rooflights, should be provided for. Coverage of between 10 and 15 per cent of the roof space typically provides the optimum amount of daylighting in a cost-effective manner. Above this level, additional costs are incurred without adding significantly to the level of daylighting provided. In addition, access to the roofs and windows will need to be considered to ensure that the windows are maintained and kept clean. The reduction in daylight through windows and rooflights that are dirty is significant, and is likely to result in increased use of artificial lighting. Consequently, it is important to model the lighting levels around the perimeter to avoid too much artificial lighting and to reduce the conflict with daylight.

Improving the efficiency of the lighting installation is one of the most cost-effective measures that can be implemented. When lamp replacement is necessary, installing more energy-efficient lamps is key to achieving a quick gain because they are only marginally more expensive than the less efficient type.

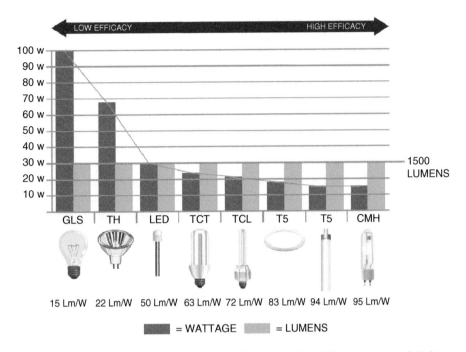

Figure 8.5: Differences in the wattage and lumens for different types of lighting. Source: Low Carbon Fit out Guide, Land Securities.

Significantly, an IRR of 26 per cent and a four-year payback can be achieved for industrial facilities, or an IRR of 42 per cent for office buildings.

This improvement is straightforward in a refurbishment situation; however, it becomes more difficult in an occupied building where work must be undertaken outside normal working hours to avoid disruption to the tenant. Out-of-hours working commands a premium, although this has little impact on the viability of this improvement compared to other energy-efficiency measures – in offices, the IRR drops to a minimum of 27 per cent.

The challenge with regard to retail buildings, however, is the retailer's preference for a particular lighting level to increase sales rather than to become more energy-efficient.

In the US, there are still more than one-billion T12 lamps in operation that use magnetic ballasts. Ballast replacement should also be considered as part of any lighting retro-fit. Replacing magnetic ballasts with basic electronic ballasts can save an additional 10 per cent of energy consumption, with an even greater saving when premium electronic ballasts are used. The change to electronic ballasts can achieve an almost 50 per cent reduction in energy usage. Ballasts can also be upgraded, which can include instant-start ballasts for dimming and sensor use, or power-shed ballasts to coincide with demand–response programmes.

The provision of effective lighting controls and systems is necessary to maintain the required lighting levels and to avoid conflict with the increased daylighting available or with unoccupied space. Artificial lighting can be automatically controlled through daylight sensors, motion detectors and zoned controls in order to switch off lighting in unoccupied areas – this could cut lighting costs by 15 per cent. In addition, internal and external time controls can be provided to align with occupancy and lighting-up times, together with passive-infra-red (PIR) and photo-electric-control strategies.

The use of light-emitting diodes (LEDs) in recent years has been driven by a number of benefits that the LED provides. In particular, there are the energy savings generated, as the LEDs are better at converting energy to light and, therefore, a lower wattage is required to provide the same output. A by-product of this benefit is the reduced heat load that a building needs to cope with, which is particularly important with increasing summer temperatures placing an increased demand on cooling equipment. They also last much longer – up to 70,000 hours (8 years) and, therefore, require less maintenance and down time during re-lamping operations.

Air-Conditioning Fan-Coil Units

Where buildings have cooling systems, energy-efficient, fan-coil units are another key efficiency improvement. Fan-coil units regulate the temperature within a given space, such as an office floor. There is now a selection of fan-coil units on the market that are fitted with direct-current motors to improve their energy efficiency. The additional cost of these higher-efficiency units, compared with that of conventional units, is small, which enables an IRR in the region of 40 per cent to be achieved with a three-year payback period at current energy prices.

However, replacing fan-coil units is not possible where a building is occupied due to the disruption caused to the tenant. Upgrading the fan-coil units would form part of a full-scale services update for which vacant possession is required.

Heat Recovery

A heat-recovery system employs a heat exchanger between the supply and extract airflows to reduce the amount of energy lost through the exhausted air, whether heated or cooled. The system is only applicable to air-conditioned and ventilated buildings. Heat recovery ranks in the top-six of improvements for offices and provides an IRR in the region of 16 per cent in most cases. For supermarkets, on the other hand, the improvement ranks only in eighth place and the energy savings are not sufficient to pay back the initial capital costs over the life of the asset and, therefore, it does not provide a positive IRR. A heat-recovery unit can be fitted retrospectively to an air-conditioning system, however, the system needs to be off-line to allow this to take place, which means out-of-hours working with an associated cost premium. The effect is to reduce the IRR to 8 per cent for the majority of offices.

Condensing Boiler

When replacing an older boiler, the sizing of the unit and the numbers involved are critical. It is usual for a number of boilers to be planned, not only as a contingency should one boiler go down, but also to ensure that efficiencies are maintained whereby only the necessary heat and hot water are provided. In order to achieve the optimum sizing and benefits, hot-water usage should be assessed with a view to reducing consumption, such as that in kitchen areas. Pipework and storage tanks should be insulated tand there should be a review of the size of the actual storage tank. In addition, a set of controls to provide effective management is vital.

A standard-market refurbishment typically replaces an existing boiler with one that is 85-per-cent-efficient in terms of its gas combustion. Higher-efficiency boilers are available, such as a 95-per-cent-efficiency condensing boiler, which has a marked impact on emissions for all buildings, less so for the modern office. The external fabric of the modern office is more efficient at retaining heat, which reduces the heating demand to be met by the boiler. The IRR varies across the offices analysed from 10 per cent to 16 per cent, with the exception of the mechanically-ventilated office which gives an IRR of 5 per cent. A 95-per-cent condensing boiler is ranked fourth in the list of the most cost-effective improvements that can be implemented in the case of supermarkets, however, the IRR is low at only 1 per cent.

The landlord of a heated industrial/warehouse building may choose to replace existing gas-fired heaters with a more efficient type. This improvement is ranked second, behind that of improving lighting efficiency, and gives an IRR of 8 per cent and payback after 11 years. However, the current preference is to let industrial and warehouse units without lighting or heating, so that it is left to the tenant to subsequently install lighting and heating units to suit their own requirements. Once again, the replacement of a boiler can be carried out during normal working hours when a tenant is occupying the building. This is with the proviso that the replacement is undertaken in summer and that there is a standby boiler to meet domestic hot-water demands.

Heating and Hot-Water Systems

Reducing the energy consumption from heating and hot water can be a very cost-effective measure, particularly where the heating system is to be upgraded. Further improvements in energy efficiency can be achieved where the building is insulated prior to the replacement or upgrading of the heating and hot-water systems. Replacement of the heating and hot-water systems is commonly controlled by country-specific building regulations. The design of an energy-efficient heating and hot-water system incorporates the following points:

- The system is correctly sized, taking account of heat gains from the sun, people, lights and appliances;
- It ensures that pipe runs are effectively insulated and that pipe sizes are increased where appropriate, leading to pumps needing to work harder;
- It uses fuel as efficiently as possible;
- It provides heating and hot water, only when required and where needed;
- It has controls that are easy to use and to understand.

An efficient heating system will reduce running costs and can be seen to add value to a property. A complete-system replacement provides opportunities for improving the energy efficiency and for assessing the option of which heating fuel to use. A partial replacement, such as that of replacing a boiler, still affords opportunities for improving the standards of controls and insulation.

Power-Factor Correction

Installing a power-factor correction unit to achieve a factor of 0.95 is also a key improvement. A power-factor correction unit reduces the transmission losses of an electrical circuit and, therefore, reduces the amount of electricity required to meet a given demand. Power-factor correction generates an IRR for offices mostly between 9 and 12 per cent with a payback within 15 years. However, the IRR is lower for an office without air-conditioning at 5 per cent. Installing a power-factor correction unit is less significant for the supermarket and industrial unit and does not provide a positive IRR for these buildings.

The power-factor correction improvement can be undertaken during a lease period although the work needs to be carried out during weekends or evenings as it requires a temporary shutdown of power. Consequently, the IRR for undertaking this work out-of-hours in office buildings falls to approximately 4 per cent.

High-Efficiency Chiller Units

For the post-2002 office, where cooling accounts for a larger proportion of energy consumption compared to that of the older offices, installing a very efficient chiller unit ranked as fifth in the list of key improvements. A chiller with a coefficient of performance (an efficiency ratio) of 11 was modelled in this scenario. The energy savings were not sufficient, however, to give a positive IRR. A high-efficiency chiller is also appropriate for a supermarket. Replacing a chiller in an occupied building can be carried out at weekends although it is a significant task with potential access constraints, and its integration with existing building services also needs to be considered.

Control Systems

Any part of the building that is unused or rarely occupied is likely still being heated, cooled and ventilated and this causes a potential waste of energy. Typically, in offices, workplace utilisation is around 60 per cent, which can lead to significant opportunities to reduce the energy demand.

The role of control systems is to match the internal environmental conditions with those related to the building occupation. This matching takes place over time and will vary through the seasons and the day; it will therefore be in constant flux rather than be a standardised pre-programmed absolute. The controls will also need to cope with those areas that are infrequently occupied, to determine when the area is used, either through occupant detection or through manual-user control.

For centrally-controlled systems, time programming and zoning are critical components. An enhanced time programme is necessary to enable the system to cope with a level of user flexibility, to change settings and the use of space. This is coupled with the ability to provide rapid heating and cooling changes when areas are not occupied, enabling the system to effectively switch off in these areas.

This is also useful in determining the heating and cooling requirements arising from the difference between internal and external temperatures.

The second aspect of centrally-controlled systems is zoning, which will help to achieve the objectives identified above. For occupants at the local level, the controls must be easy to understand and use, and be located in an appropriate position at the point of need. In addition, a feedback loop must be available to address concerns and meet requirements that are outside the scope of the standard controls on the control panel.

The management of the control system at a central level is commonly provided by a Building Management System (BMS), which can provide control over the heating, cooling, lighting and air-quality systems. Modern systems can also 'talk' to a range of other systems, including the security systems to aid detection of individuals within a building, and to ensure that the necessary internal, working, environmental conditions are provided.

Unfortunately, many systems are poorly used, overly complex, not well understood by the operator and do not effectively manage the internal environment, leading to a significant level of energy waste. The BMS is a live system and will need to vary based on the business activities and functions; it is therefore a dynamic system. It is important to understand how the system will operate and to know who will be looking after the BMS, so as to specify the correct version and to ensure that skilled personnel are available.

Energy efficiency in the Middle East and North Africa[11]

The barriers to achieving a low-carbon situation and energy efficiency in the Middle East and North Africa are no different from those in any other country, with knowledge gaps, lack of financing and split incentives being the main obstacles.

Whilst the study identified an energy-saving potential of 57 per cent across the ten pilot projects, the payback periods were commonly between ten and twenty years, with only one project being below this level. Paybacks of up to 60 years were identified for some of the projects. In some cases, this was due to subsidised fuel, such as in Algeria, where a 57 per cent reduction in energy equated to a cost saving of €70 per year. In all projects, high costs were seen to be necessary to deliver the projects due to the uncertainty, high level of perceived risks and high costs associated with the technology.

The three main barriers identified were:

1. Knowledge gaps – clients are not aware of the technical and financial potential; developers lack experience to procure and implement the technologies; and higher operational costs are incurred from the monitoring and training of the relevant staff.

(Continued)

2. Lack of financing from clients to accept the upfront costs directly, with banks being unwilling to provide investment.
3. Split incentives arise from the diversified supply chain, which has its own drivers and different risk levels. Critically, the landlord–tenant relationship was seen as being a major issue resulting from the increasing commercial model.

The conclusion was to request government intervention to remedy the market failure surrounding the high cost of technologies and the subsidies for energy. Policies addressing taxes, financial incentives and promotion through campaigns were seen as the most viable way forward.

8.6   ESCOs and Energy-Performance Contracting

There are a number of financial incentives and tax breaks available through government funds and bonds to promote energy-efficiency measures. These are typically limited to specific activities or capped at a certain value above which the investor has to pay directly. Given the current state of the financial markets, the ability of banks to provide the necessary investment can be limited and, as a result, the lack of available cash might cause some viable projects to be put on hold.

An increasing trend is to seek funding through an Energy Service Company (ESCO), which can provide the upfront capital investment in return for planning, installing and operating the energy-efficient equipment for a fixed period of time. Equipment can include many of the mechanical systems described in Part 8.5, such as the control systems and operational-management systems. Savings are generated from the improved efficiencies, thus providing the ESCO with a return on their investment.

The Energy Performance Contract (EPC) is the vehicle most commonly used to finance the energy retro-fits. The use of EPCs is common in the US and Australia, and also across much of central Europe, particularly France, Germany and Austria who all have a long-term involvement with ESCO projects[12]. Performance-based contracting means that the compensation and project financing for a company that undertakes a project is directly linked to the amount of energy that is actually saved when the project is complete[13].

Typical features of Energy Performance Contracts include[14]:

- Guaranteed Savings: The most common type available in the US and Australia, whereby the level of energy savings is guaranteed and compensates the property owner if the levels are not achieved.
- Shared Savings: The most common type used in Germany, whereby energy savings are shared between the property owner and contractor for a set period of time. The contractor is usually responsible for financing the project.

- First Out: All savings achieved belong to the contractor until their costs that were incurred in the project have been paid, whereupon the contract ends. After this point, any further savings achieved belong solely to the property owner.
- Chauffage: The contractor charges a fixed fee and, in return, the property owner receives a guaranteed level of service. The contractor is responsible for procuring the energy, paying bills, carrying out operations and maintenance. The equipment supplied by the contractor will belong to the property owner at the end of the contract.
- Energy Supply: Contractors receive a fee for creating energy as a by-product from existing business processes, for example, through co-generation.

Energy Service Company

The intention of the ESCO is to procure energy, make efficiency savings, or generate reductions in carbon dioxide emissions. In simple terms, the characteristics are:

- It guarantees the energy savings and/or provision of the same level of energy service at lower cost;
- Its remuneration is directly related to the energy savings achieved;
- It can either finance, or assist in arranging financing for, the installation of an energy project that it implements by providing a savings guarantee.

ESCOs are generally, although not specifically, special-purpose vehicles (SPVs) and can incorporate some, or all, aspects of the design, build, finance, ownership and operation of energy assets, through to the provision of energy-efficiency advice and the installation of energy-efficiency measures. The same ESCO can be responsible for both energy-generating activities and energy-saving activities. In some instances, investment can be provided by specialist companies, depending on the size of the project and on whether the investment returns are sufficient.

Descriptions of ESCOs can fall into the following groupings:

- Public-sector or private-sector driven, and operated on energy-performance contracting principles;
- Private-sector driven, with (or without) public-sector encouragement;
- Public-sector driven, but procured and operated by the private sector, though not fully on energy-performance contracting principles;
- Public-sector driven with no (or very little) private-sector involvement;
- Public-sector driven with private-sector involvement in design and build.

(Continued)

The decision as to what sort of project to pursue, and the particular structure to set up, will depend on project-specific matters, which include costs; the attitude to funding; the attitude to risk; the degree of knowledge and experience within the contracting body on energy matters; whether the project is concerned with the construction and operation of a single facility or whether it is intended to cover a range of energy-efficiency matters; and so on.

In addition to deciding on the project type, the financing or 'energy performance contracting' (EPC) model also needs to be considered. There are a number of possible models where the capital-improvement budget allows for the funding of efficiency upgrades through cost reduction. These include options for both shared savings and guaranteed savings.

ESCO Operation

An ESCO (Energy Service Company) is a company that designs, installs and operates energy services over periods of up to 30 years. Financing a project through an ESCO can therefore transfer the capital costs and risk of the project to a third party. An ESCO can typically carry out the following services:

- Design;
- Finance;
- Installation;
- Operation & Maintenance;
- Monitoring and Emissions' Reporting;
- Customer Services and Billing.

The level of risk transferred to the ESCO will depend on the set-up of the project and the ESCO itself. The setting up of a special-purpose com-pany, specifically for the delivery of an individual project, will allow investment from interested parties (e.g. the developer or council) and the associated return on investment for these parties. Alternatively, the ESCO provider can have full ownership of the project and thus accept all the financial risks.

An ESCO will typically undertake a community-energy scheme over the course of 20–30 years. This will allow the capital costs of setting up the community-energy scheme to be recouped and profit to be accrued over the course of the project. The ESCO will do this by running the community-energy scheme, providing operation and maintenance to the project and selling energy to customers. The ESCO will purchase the fuel required to run any equipment servicing the project and will sell heat and electricity to the customers of the scheme. The prices of heat and electricity to con-sumers are usually guaranteed to be below the comparative prices charged by the leading energy suppliers in the area.

The benefits to customers of buying heat and electricity from a community-energy scheme supplied by an ESCO are:

- No maintenance costs associated with individual boilers;
- Dwellings do not have discrete boilers;
- Prices of heat and power are guaranteed to be lower than the cost from other local suppliers for the term of the project.

The main benefit of selecting an ESCO to supply energy services to a site is the transfer of capital costs and financial risk to a third party.

Customer Metering and Billing

Billing and customer care can be part of the services provided by the Energy Service Company (ESCO) and can be incorporated into the site-development scheme, as can returns from incentives.

The rule of thumb for renovating buildings is that it is relatively easy and cost-effective to reduce energy consumption by 30 per cent in many older, commercial buildings. In fact, energy-management companies will often guarantee specific savings, such as 28 or 32 per cent, as part of a performance-management contract, and will pay the difference if that level of savings is not achieved. These energy-service companies also will secure funding for the upfront costs and accept repayment exclusively from all, or a portion, of the money saved through lower energy bills.

For an energy-management contract to work, the building owner and the ESCO need to establish a baseline. How much energy and electricity is the building currently using? There will be some contractual complexities, such as distinguishing between those future savings that are due to greater efficiency, those that occur because a tenant moves out, and those due to unusual weather. However, these are quite common issues that have been dealt with many times previously. The key is to simplify the contract to make it understandable; this will help to reduce its complexity and convince the owner of the building to purchase an efficient building-energy package.

If the energy-management company is going to guarantee the savings, it needs to have sufficient control to ensure that the equipment in the building is well maintained and operating in the most efficient way. ESCOs also are concerned about tenant behaviour in so far as their activities contribute to the level of savings envisaged within the Energy Performance Contract. To achieve significant energy savings, tenants need to cooperate by turning off lights as appropriate and by keeping the temperature in their particular space at a reasonable level.

A number of lessons have been learned from previous projects, including that of the Empire State Building project[15], as to how to get the best from ESCOs and Energy Performance Contracts.

1. **Taking a comprehensive approach:** Most energy-service companies (ESCOs) do not capture all available energy gains. This is because they rarely take a comprehensive approach and building owners do not specifically ask for a 'whole-building' analysis. Hence, most projects focus exclusively on meeting loads efficiently, instead of reducing them or eliminating them all together. This often leads to retro-fit projects that generate about 10 per cent to 15 per

cent savings from easy and quick payback measures, such as improved lighting and HVAC controls. By considering energy-efficiency measures that reduce energy loads, such as building-envelope changes and daylighting changes, in addition to installing more efficient systems to meet existing loads (e.g. lighting and HVAC), a whole-building retro-fit can achieve much greater energy savings whilst improving the project economics.

2. **Aligning energy retro-fits with equipment replacements:** For an energy-efficiency retro-fit to be cost-effective, the retro-fit needs to be aligned with the planned replacement, or upgrades, of multiple building systems and components. For instance, the Empire State Building had plans under way to replace its chillers, fix and reseal some of its windows, change corridor lighting and install new tenant lighting as leases expired. Since these upgrades were already on the books, the design team redesigned, eliminated and created projects that cost more than the initial budget but would generate significant energy savings over a 15-year period. When these energy savings were taken into account, along with the added upfront project costs, the net present value of the energy-efficient retro-fit projects was better than that of the initial retro-fit projects. The key message to take away here is that you actually have to align retro-fits with the scheduled replacement of equipment; otherwise, the energy savings would typically not be substantial enough to offset the full, non-incremental, capital cost and could justify only the incremental, capital cost (e.g. the difference in cost between resealing windows and re-manufacturing windows).

3. **Capturing tenant energy savings:** In existing commercial buildings, in order to take advantage of the full energy-efficiency opportunity, it is necessary to engage with the tenants. In the Empire State Building retro-fit, more than 50 per cent of the energy savings were achieved only with some level of tenant engagement. Often energy-efficiency opportunities are not grasped because of numerous real, or perceived, barriers, including a hesitation to engage with tenants, split incentives, business-interruption concerns and so on.

4. **Finding a financing model that works:** Financing building retro-fits through private capital alone will be difficult in today's lending environment. Given tight lending conditions, it is unlikely that private capital will be widely available for retro-fits of privately owned commercial buildings. In addition, since most first-mortgage liens cover existing equipment that will be replaced or upgraded in a retro-fit, an inability to collateralise loans might further complicate the financing situation. A number of possible solutions that combine private capital with publicly funded loan guarantees and other public financing mechanisms have been proposed, but no single financing mechanism has gained national traction.

5. **Retro-fitting small and medium-sized buildings:** The ESCO business model, employed mostly for the retro-fit of municipal, state, federal, university, school and hospital buildings, and some large, commercial-office buildings, is not suited to small and medium-sized, commercial buildings, where transaction costs are often disproportionately high relative to the profit opportunity for ESCOs. Small buildings with less than 50,000 square feet of floor space account for approximately 50 per cent of the total commercial-building stock in the United States. This presents a major opportunity for business-model innovation to deliver retro-fits to a large and under-served commercial-building market segment.

Clinton Climate Initiative's Energy-efficiency, Building Retro-fit Programme

The programme brings together building owners, cities, energy-service and technology firms, and financial providers in an effort to reduce energy consumption in existing buildings. CCI is working on over 250 projects, encompassing over 500 million square feet of building space in more than 20 cities.

1. **Energy-efficiency financing:** to address lack of collateral or credit, owner–landlord split incentive, and transfer or sale of buildings.
2. **Energy-performance contracting (EPC):** where an energy-service contractor provides project design, implementation and measurement, along with a financial-performance guarantee.
3. **Technical retro-commissioning (RCx): where the** focus is to be placed on the optimisation of existing building equipment and operations, rather than on the replacement of major equipment.
4. **Energy-efficient technology procurement: to** develop procurement models to assess product options from a life-cycle perspective rather than on a 'lowest first cost' basis.

K. Raheja Corp (KRC), Mumbai, India

In 2007, KRC, a leading real-estate corporation that owns many properties in India, began working on three retro-fit projects in Mumbai: the Inorbit Mall Malad, The Resort Hotel, and Hotel Renaissance.

An EPC approach was used, focusing on improvements in building systems and upgrading equipment.

- HVAC upgrades;
- Chiller-plant improvements;
- Hot-water upgrades;
- Lighting improvements including LEDs;
- Sewage treatment-plant improvements;
- Solar-powered LCD 'dashboards'.

KRC faced the challenge of obtaining both management approval and shareholders' consent. No premium from tenants and building occupants would be received for implementing energy-saving measures, requiring approval for the work as a capital investment on a 2.5-year simple payback.

As energy-efficiency retro-fitting was a new concept in India, training for architects, engineers and building managers was required – adding extra time and expense to the first project, with subsequent projects being able to implement the lessons learnt.

8.7 Energy-Efficiency and Low-Carbon Checklist

Table 8.5 provides a summary of the items to be discussed during a sustainable-refurbishment project, capturing where specific items are necessary, recommended or nice-to-have options.

Table 8.5: Energy-efficiency and low-carbon checklist

Minimum Standard	Recommended	Nice to have
Design energy-efficiency to meet industry standards and local building standards	Design energy-efficient measures to meet best-practice industry standards	
Check energy-performance indicators 12 months after occupation against the original design criteria	Target annual 10 per cent reduction in energy performance during operation	
Install a permanent carbon dioxide monitoring system linked to BMS and specify initial set points		
Procure business equipment (MFDs, computers etc.) with high-efficiency, energy-management systems; check quiescent power-down is set	Minimise purchasing of electrical equipment through increased multi-functional use, and use of electronic media	
Install easily understood labels to turn off lights when not in use	Install occupancy sensors in areas with intermittent usage	
Where macro/microclimate and site permit, consider use of natural energy sources (e.g. ventilation through opening windows, groundwater heating/cooling etc.) and 'mixed mode' operation	Assess use of chilled beams and increased thermal mass to provide natural heating and cooling in the building (optimise building envelope)	Procure 'green electricity' (i.e. electricity from sustainable sources)
Assess the feasibility of renewable-energy generation – PV, photo-thermal, wind, CHP	Demonstrate that design includes means to retro-fit (further) green or 'low carbon' energy equipment in future – future-proofed against impending legislation and future technologies	
Specify a BMS linked to site operations, e.g. lighting, plant	Assess use of night cooling – ice-storage chillers to minimise daytime energy consumption	

8.8 Health and Indoor-Environment Checklist

Table 8.6 provides a summary of the items to be discussed during a sustainable-refurbishment project, capturing where specific items are necessary, recommended or nice-to-have options.

Table 8.6: Health and indoor-environment checklist

Minimum Standard	Recommended	Nice to have
Install daylight responsive controls within 4.5 metres of windows; avoid meeting rooms and offices on the perimeter – design for open plan		
Meet, and surpass, local building standards for ventilation	Assess viability of natural ventilation; utilise thermal modelling to determine optimum profile	Assess viability for heat-recovery ventilation
Select/specify/design buildings with good daylight potential; target minimum 3–5 per cent daylighting factor in spaces 4.5m from perimeter walls, with good DF uniformity and adequate daylight over 80 per cent of office space	Install reflectors or provide an atrium space to increase daylighting throughout the building	
Provide window blinds/(insulated) film on windows to reduce glare and discomforting solar heat radiation; locate workstations away from the windows	Provide baffles to minimise glare and heat gain	
Install automatic-occupancy lighting control; use photocell control where daylight potential allows; consider low-background luminance with local-timed task lighting	Install high-frequency ballasts in all office areas	
Comply with industry standards and local building standards for clean air and ventilation; plan and conduct a 1-week building flush-out after construction and prior to occupancy	Provide for air-change effectiveness greater than, or equal to, recommendations; demonstrate air-flow pattern that involves greater than 90 per cent of the room or zone, and a ventilation effective of not less than 1.0	Undertake thermal modelling to optimise thermal comfort
Determine the possible use of LED lighting and sunlight pipes in back-of-house areas	Direct line of sight to external areas through glazing from 90 per cent of all regularly occupied spaces	

Endnotes

1 The Non-Domestic Buildings Refurbishment Report from Caleb Management Services (February 2011)
2 http://www.institutebe.com/Energy-Efficiency-Indicator.aspx
3 WGIII/ SPM, p. 13
4 The Probe Study 2001, Building Research and Information, vol. 29, 2 March 2001
5 AIA's 50 to 50 Programme is available at www.org/fiftytofifty
6 Costing Energy-efficiency Improvements in Existing Commercial Buildings – summary report Jan 2009, IPF
7 Costing Energy-efficiency Improvements in Existing Commercial Buildings – summary report Jan 2009, IPF
8 Costing Energy-efficiency Improvements in Existing Commercial Buildings – summary report Jan 2009, IPF
9 The handbook of Sustainable Refurbishment – Non domestic buildings, Nick V Baker, Earthscan 2007
10 Lawrence Berkeley National Laboratory, Heat Island Projects, http://eetd.lbl.gov/heatisland/PROJECTS
11 www.med-enec.com
12 Developments of Energy Services Companies across Europe: A European ESCO Overview, Wilhemus de Wit; European Commission, DG TREN; www.rusrec.ru/files/11_De_Wit_ESCO_ENG.ppt#435
13 What is an ESCO?, National Association of Energy Service Companies (NAESCO), www.naesco.org/resources/esco.htm
14 Adapted from 'Energy Performance Contracting: Ways to reduce Greenhouse Gas Emissions for Small and Medium Enterprises'; Australian Government, Department of Environment, Water, Heritage and the Arts; www.environment.gov.au/settlements/challenge/publications/factsheets/fs-energy-performance.html
14 Details taken from www.esbsustainability.com and www.esbnyc.com

9 Behavioural Change

The undertaking of a refurbishment, particularly at the medium-level scale where tenants or occupants may still be in situ, can provide environmental and social benefits to the occupants resulting in significant cost savings. Typical benefits include not only improved energy efficiency, but also a reduction in absenteeism and an increase in productivity. These benefits derive from the measures taken during the refurbishment but they can quickly diminish if their development is neglected.

Whilst the refurbishment may have identified a number of areas in which to significantly reduce carbon emissions, if the employees working in the building are not made aware of the situation, then the savings may not be realised. This also applies to the technical personnel, facility managers, operational staff and the other managers, who all have a role to play in achieving an environmentally efficient building.

Whilst energy-efficient building design will help the occupants reduce their energy consumption, the actual usage of energy is still largely dependent on the aggregate of the individuals' behaviour within the building. This can be difficult to control within a commercial environment because overall energy consumption is not always easy to attribute to specific individuals or building areas.

Although improved monitoring, procedures and management systems can significantly reduce consumption, behaviour of individuals is often driven by an organisation's ethics rather than by an awareness of the economics, or environmental impact, of energy consumption.

Control of energy use within smaller buildings is potentially easier to exert because responsibilities are clearer; there are some quick gains to be made by introducing behavioural changes or by upgrading the efficiency of some of the building elements. However, many of the more substantial savings require larger

Sustainable Refurbishment, First Edition. Sunil Shah.
© 2012 Sunil Shah. Published 2012 by Blackwell Publishing Ltd.

investments. Payback periods can be long – and potentially extend beyond the intended period of occupancy. Furthermore, domestic properties are 'homes' and thus decisions are often based on personal preference rather than on efficient functionality.

There are, however, signs of a new emphasis. The introduction of Display Energy Certificates as a mandatory requirement across Europe and parts of the US, whereby actual energy usage is reviewed and presented, suggests that stakeholder interest has been recognised as potentially the most potent motivator for control of energy – at an organisational level, at least.

Chapter Learning Guide

This chapter will look at the role that behavioural change can play in maintaining the benefits achieved from a sustainable refurbishment.

- Commissioning the refurbished area and transferring the necessary knowledge to the operators of the building;
- Maintaining employee behavioural change through a specific project.

Key messages include:

- Buildings do not use energy – it is the people within them that do;
- Understanding the building, and changing the behaviour of the employees working in the building, can reduce energy use and absenteeism, together with improving productivity.

9.1   Commissioning Buildings

One of the most cost-effective ways to overcome behavioural 'discrepancies' is through the process of commissioning a building. Only about 1 per cent of buildings are commissioned, according to the US Department of Energy, probably because most building owners are wary of the upfront cost of commissioning and also the cost of fixing the problems that might be identified in the process. A review of published and unpublished data on 224 buildings in 21 US states, representing 30.4 million square feet of commissioned space (73 per cent in existing buildings, 27 per cent in new ones), was undertaken to analyse the costs and benefits of commissioning[1].

Total commissioning costs for these buildings were US$17 million (2003), an average of US$0.55 per square foot. Amongst other findings, the review revealed that:

- An average 11 deficiencies were found in existing buildings, 28 in new buildings. HVAC systems represented the bulk of the problems.

Table 9.1: Typical deficiencies found during the commissioning of school buildings[2]

Excessive play or gap in dampers	Malfunctioning power exhausts
Inoperative dampers and actuators	Malfunctioning economiser controls
Incorrect programming of building-use patterns	Oversized fans, pumps and related equipment
Unapproved field modifications	Direct-wired exhaust fans always on or malfunctioning
Dirty filters and coils	Improper set points
Water leakage on electrical equipment	Improper CO_2-based purge operation
Improper flue exhaust	Incorrect commissioning of renewable technologies

- For existing buildings, median commissioning costs were US$0.27 per square foot; energy savings came to a median 15 per cent (18 per cent average); payback times were lower than nine months (0.7 years).
- For new buildings, commissioning costs were US$1.00 per square foot (0.6 per cent of total construction costs), yielding a median payback of 4.8 years.
- Reduced change orders and other non-energy benefits accounted for savings of US$0.18 per square foot in existing buildings and US$1.24 per square foot for new construction – for new buildings, enough to cover the entire cost of commissioning.

The lack of an effective commissioning programme can reduce, and certainly undermine, the benefits achieved through the refurbishment, whether for a leased property or for an owned property. Whilst not a cure-all solution, it is one of the most cost-effective and far-reaching means of improving the energy efficiency of buildings. Table 9.1 provides some typical examples of deficiencies found during the commissioning of school buildings.

Why aren't more building owners taking advantage of commissioning? One reason is inertia. Many building owners just accept higher energy costs as a fact of life – and either absorb them or pass them on to their tenants. The fact that only 45 public-sector companies of BOMA's 16,500 members have taken up the 7-Point Challenge is a sign that building owners would rather live with the problem than address it. Building size is another limiting factor. According to the USDOE's Energy Information Administration, 98 per cent of commercial buildings in the US are less than 100,000 square feet in area. They comprise about two-thirds of the total floor area and consume about 60 per cent of the energy used by buildings in the US. Since the 'fixed costs' (mostly labour) of hiring a commissioning-resource provider are roughly the same, regardless of building size, the cost of retro-commissioning smaller buildings – estimated at US$0.40 to 0.60 per square foot – is greater than that for large buildings (US$0.27 per square foot for the median 151,000-square-foot building [3] [4].

Many owners, even the enlightened ones who have commissioned their buildings, fall into the trap of thinking that it's a one-off event. In fact, building systems, particularly HVAC systems, are forever falling 'out of tune', even in new buildings. This raises the question of the need for more and better training of facilities' personnel to enable them to carry out the commissioning on a day-to-day basis, as well as the need for periodic (some even advocate 'continuous') recommissioning. This should be part of a behavioural change for the management and operation of buildings.

There are a number of areas where energy is wasted in buildings; energy wastage can commonly be managed and reduced through general awareness, resetting of controls and implementing lower levels of refurbishment. Reasons for wastage include:

- Equipment running more than needed;
- Cooling or heating air more than needed;
- Cooling or heating water more than needed;
- Heating and cooling running at the same time;
- Moving too much air;
- Moving too much water.

Mistakes abound even in the newest buildings:

- Fans in air-handling units running backwards;
- Temperature sensors placed in direct sunlight, making their readings inaccurate and unreliable;
- Vibration-isolation components in the shipping position instead of in the operating position;
- Missing gauges;
- Setpoints not inputted.

9.2   Energy Conservation as a Behaviour

Leadership is essential, and there are many inspiring examples where ambition has been translated into action, typically at the technological, systems and policy levels, rather than at the level of the day-to-day behaviour of employees. Very few companies have truly nailed the behavioural-change model that translates the changing of business values into daily action. Truly embedding sustainable, behavioural change requires a combination of engagement techniques, training and reward.

Many companies have realised that recruiting and supporting 'environment champions' is one of the most effective ways to drive positive change. But too often these initiatives appear as a bolt-on to existing personnel policies, thus increasing their chances of failure. Volunteer champions who put their name forward in good faith can find themselves in all ill-defined role, for an unspecified period of time, with little relevant training, with limited support from their line manager and knowing that their efforts will not necessarily be recognised in performance reviews and bonuses.

There is no doubt that environment-champion initiatives can be one of the most effective ways to bridge the gap between values and actions, but such initiatives do need to be carefully planned and managed. They must be integrated with personnel policies and fully backed by the business, otherwise there is a real danger that the most enthusiastic and dedicated employees capable of driving the change will be hung out to dry.

How to Get Energy Conservation into Your Workplace

1 KNOW YOUR ENERGY PROFILE
Your energy profile is your energy consumption pattern based on half-hourly meter readings. In many businesses the profile resembles a top hat, with the highest consumption being from late morning to mid-afternoon. Your organisation's profile may be completely different, but it doesn't matter as long as you know what it looks like. By understanding your profile, in a couple of days you'll be noticing when and where unusual consumption occurs.

2 CREATE A COMPELLING MESSAGE
Successful energy-conservation projects start with a compelling message. It has to be more than just 'saving money' or 'reducing carbon', which will have little appeal and make everyone feel guilty. Consider something aspirational such as a commitment to ISO 14001 or a Green Policy commitment to your customers. Think of other successful campaigns and copy them.

3 BEHAVIOUR IS EVERYTHING
Your challenge is to engage and empower your colleagues in reducing energy consumption. The good news is that they really want to change; the bad news is that they fear they won't sustain it. To help that change, you'll need to persuade them that energy conservation is one of your organisation's Key Performance Indicators (KPIs). Having conservation KPIs will not only sustain their focus but also reward them with improved profits.

4 IDENTIFY THE INFLUENCERS
Identify the 'connected' people in your business. Don't assume that they are the senior managers. Often, corporate initiatives are viewed with cynicism. Conspire to involve good communicators such as receptionists, PAs, section heads, team leaders, and sales and marketing people to act as your ambassadors.

5 CREATE ACTION PLANS
Once you have the team, get together and list out the areas and actions where you believe you can make the biggest savings fastest – the premier division. Then list out the first division – the areas and actions where you believe you can still make savings but where the impact will be less great and less rapid. Involve the team members to communicate details of the areas and actions to their respective audiences.

6 COMMUNICATE SUCCESS
Resulting from many small behavioural changes, the actions of your colleagues will very quickly accumulate to create the savings you targeted. To acknowledge their contributions, your action plan should include a clear

(Continued)

and frequent communication strategy. It is recommended that weekly consumption graphs against KPIs are displayed on notice boards.

7 KEEP IT CLEAR AND SIMPLE
Only energy-industry and climate-change bores think that talking about CO_2, M3, F3, KVA and so on is cool. When you're discussing energy conservation, keep language and targets simple.

8 COMMUNICATE THAT ENERGY CONSERVATION IS A LONG-TERM, GOOD BUSINESS DECISION
Think of energy conservation as 'the third energy source', after those of fossil fuels and renewable fuels. This energy source is also an excellent, long-term contributor to your business, because it has little cost but the potential for significant savings through reducing energy consumption, thereby helping to provide increased profit to the business.

In order to effectively build behavioural change into any refurbishment project around the knowledge – empowerment – feedback framework, the following measures are necessary:

- Making it easier for people to be involved and be supported – any complexities will result in individuals' finding reasons not to be engaged;
- Tailoring the feedback so as to be comparative, and also location-specific to those who are being targeted. However, feedback on performance alone is likely to provide only a short-term change, perhaps lasting a few months, without delivering a longer-lasting change;
- How to build energy conservation into the organisation's culture is critical – the provision of information by itself will not be sufficient to change the culture;
- A competitive element involving incentives, which can be fun, will help to engage people as part of an overarching programme, with a number of activities taking place;
- Having the sustainability option as the default option, for example, installing delayed-opening lift doors to encourage walking, which has a beneficial impact.

Ten key principles to be applied include:

1. People use energy, buildings don't.
2. Facilities' management needs to work more synergistically with building occupiers and designers across all sectors.
3. There is a need for clearer, pragmatic thought and leadership amongst key stakeholders, with the identification of clear, long-term, adjustable goals.
4. One size does not fit all. Approaches to behavioural change and the method of disseminating learning to a company's stakeholders need to be tailored and realistic.
5. Building performance needs to be compared to organisational performance.

6. Benchmarking has a role to play, but the resulting data needs to be used effectively to shape behaviour.
7. Do not rely solely on 'eco-bling' (technology) as this creates cynicism amongst staff.
8. There is an assumption that behavioural change has no cost, indicating a need for the development of a new financial model to support this assumption.
9. 'Sustainability' and 'carbon' (the S and C words) do not resonate with many audiences.
10. People need to be applauded and rewarded for their efforts.

Energy Trophy Programme – Europe[5]

Energy Trophy (ET) is an EU-wide programme sponsored by the European Commission's Intelligent Energy Europe (IEE) Programme, the German Federal Environmental Agency and Ecoperl. The programme takes the form of a competition for energy savings, specifically in office buildings, resulting from the change in employee behaviour. The goal of the programme is to reduce energy consumption specifically through behavioural change within organisations, which does not require investment into energy-saving devices; however, low-cost investments, such as compact, fluorescent light bulbs and timers, are permitted. ET also has a focus on cost savings to increase employee participation and carbon dioxide reduction.

The first round of ET was launched in 2004–05 with 38 participating companies and institutions from six countries. The results from the first round were promising, with an average energy saving of 7 per cent and the winning company recording a 30-per-cent saving. Overall, the programme achieved a total reduction of 3,700 MWh, or 1,885 tonnes of carbon dioxide, corresponding to a cost saving of €205,000.

Overall, the ET programme is considered to be a success, as planning for future rounds of the competition is already in motion, in fact, an established framework is being developed to formalise the competition for future rounds to ensure that it runs smoothly. Furthermore, the programme offers significant benefits: public recognition through the award of prizes; publicity; corporate identification; motivation for the staff of participating organisations; and reduced operating costs as energy consumption is reduced though minimal investment.

Lessons learnt
1. Clear communication of goals and objectives.
2. Linking the objectives with the environment and multiple benefits.
3. Real gains for participants, both financial and social.
4. The full support of the EC and other sponsors that give ET a long-term perspective; its success will likely increase as the programme expands.

Endnotes

1 Mills, E., N. Bourassa, M.A. Piette, H. Friedman, T. Haasl, T. Powell and D. Claridge, *The Cost-Effectiveness of Commissioning New and Existing Commercial Buildings: Lessons from 224 Buildings'*, Proceedings of the 2006 National Conference on Building Commissioning, Lawrence Berkeley National Laboratory Report No. 56637. At: http://eetd.lbl.gov/emills/EMillspubs.html

2 Adapted from Vivke Mittal, Enovity Inc., and Mike Hammond, Folsom Cordova (Calif.); Evolution of Commissioning within a School District: Provider and Owner/Operators Perspectives; Unified School District, National Conference on Building Commissioning, 23 April 2008

3 'Think Small: The Key to Unlocking the Existing Buildings Market', Tim Kensok and Jim Crowder, AirAdvice Inc., National Conference on Building Commissioning, Newport Beach, Calif., 23 April 2008

4 A study of existing buildings >25,000 square foot by Portland Energy Conservation Inc. found that unit costs ranged from $0.32/square foot to $0.47/square foot based on average building size and depending on market sector, 'Final Report: California Commissioning Market Characterization Study', PECI, November 2000. At: http://resources.cacx.org/library/holdings/018.pdf

5 http://www.energychange.info/casestudies/165-case-study-8-energy-trophy-programme

10 Renewable Energy

A range of low-energy and renewable-energy systems can be applied to domestic and commercial buildings to improve energy performance. The application of these systems to an existing building must be considered where there is a local planning authority renewable-energy target. In addition, the 'softer' benefits of applying these technologies lie in awareness raising, both internally and externally, and in enhanced profile raising for landlords and tenants alike. The suitability of using low-energy and renewable-energy systems is very much dependant on the specific building and location under consideration, and the advantages and disadvantages of these technologies have been well documented.

Renewable technologies can be incorporated into the design of the facility. There is a level of resistance to take up certain technologies, such as micro wind turbines or photovoltaics, due to cost pressures, particularly capital-cost barriers – the long payback period combined with the high upfront investment required for the more visual solutions, such as photovoltaics, constitute a significant barrier.

Chapter Learning Guide

This chapter will highlight the commercial, low-carbon and renewable-energy technologies and their application to refurbished properties.

- Summary of technologies, including where and how they can be utilised to obtain the optimum viability;

(Continued)

Sustainable Refurbishment, First Edition. Sunil Shah.
© 2012 Sunil Shah. Published 2012 by Blackwell Publishing Ltd.

- The critical operational risk for these technologies is the correct design and commissioning of the units.

Key messages include:

- There is no single best technology, as a number of factors will affect the anticipated outputs;
- Technological development is moving rapidly. Reviewing the potential benefits regularly will help to maximise the return.

10.1 Introduction

Implementation of renewable technologies does not have to cost significantly more than traditional construction. Incorporation of the technologies into the concept design to maximise their potential does not add greatly to the capital cost. Targets for renewable energy should be set for any facility with an area of more than 1000 square metres, given that the size will justify any increase in capital cost with a short-term payback. In many cases, grants or tax concessions are available to offset capital costs, either wholly or for a period of time. The return from the carbon dioxide reduction against the investment varies greatly for the various technologies. This translates into the cost-effectiveness of the various technologies commonly available.

Solar PV costs globally have fallen from US$2 per kW in the 1970s to less than US$0.30 in 2008, although the price needs to further reduce to half of this amount in order to achieve grid parity, which is expected across many markets in the next decade[1]. In the UK, six panels, or 1 kWp, of solar PV on an average home, built to latest building regulations, will reduce carbon emissions by more than 20 per cent. Guaranteed by the manufacturers for 20 or 25 years, they will last far longer than this. In the case of building-integrated solar PV, these installations also offset the costs of other traditional roofing and facade materials. This is of particular relevance to the commercial sector. Building designs will therefore need to adapt so that they can readily integrate such systems in efficient and cost-effective ways (e.g. PV panels as a cladding material).

The chief drawback of renewable-energy sources is their cost, compared to that of conventional energy sources. The cost of generating electricity from wind is low, compared to that from solar or wave power. Although the cost of electricity from conventional sources is currently low, changes in the security of their supply and costs are enabling renewable-energy sources to compete directly. Together with the rising cost of the energy supply itself, an increasing level of taxes has been imposed on fossil-fuel energy supply to compensate for the carbon emissions and the climate-change implications resulting from the use of such an energy source. Hence, these taxes are increasing the cost of fossil-fuel energy in comparison to energy from renewable sources, helping to reduce the cost difference significantly. It is anticipated that grid parity will be achieved for many technologies in the coming decade due to these twin increases in the cost of energy supply.

Table 10.1 describes a variety of low-carbon and renewable-energy technologies available on the market, the mechanisms of how they operate and how to

Table 10.1: Summary of renewable technologies and key issues

Technology	Energy Conversion & Utilisation	Key requirements for optimising viability	Key aspects affecting implementation
Photovoltaic Panels	• Daylight to electrical energy • Enables electrical power for utilisation within the building	• Unshaded access to daylight for PV panel location • Southern orientation for PV panel • 30°–45° elevation of tilt for PV panel	• Suitability of roof structures to mount PV-panel arrays • Urban location influences potential planning restrictions
Solar Hot Water	• Solar radiation to thermal (heat) energy • Enables hot-water heating for ablutions and catering use	• Unshaded access to the sun for solar-panel location • Southern orientation for solar panel • 30°–45° elevation of tilt for PV panel	• Suitability of roof structures to mount solar-panel arrays • Urban location influences potential planning restrictions • Central provision for hot-water production required (not individual localised heaters)
Wind Turbine	• Wind to electrical energy • Enables electrical power for utilisation within the building	• Unsheltered location for micro wind-power location	• Suitable external areas to mount micro wind turbine • Urban location influences potential planning restrictions
Bio Fuel Heating	• Organic vegetable matter (tree wood) combustion converted to thermal (heat) energy • Enables heat generation for utilisation in building	• Access to sufficient, reliable local resource of renewable fuel (i.e. woodchip or pellets)	• Suitable location for bio fuel-boiler location • Suitable location for fuel storage • Sufficient provision for fuel delivery • Urban location influences potential planning restrictions (boiler flue-discharge position)
Geothermal	• Natural ground-source heat-sink conversion to higher-grade heat or cooler via electric heat pump • Enables increased efficiency in operation of mechanical heating and cooling systems (heat pump)	• Access rights for extraction of local groundwater • Geotechnical survey to establish local suitability of site	• Existing sites might not be in suitable locations to access optimum groundwater aquifers • Space for boreholes or closed-loop exchangers might be limited • If no mechanical cooling, utilisation might be limited
Combined Heat and Power	• Utilisation of waste heat from simultaneous electrical power generation at point of use • Enables localised production of heat and power – with savings in CO_2 emissions from use of waste heat	• Year-round requirement for heat utilisation	• Suitable location for CHP unit adjacent to existing boiler plant • Domestic hot-water services to be served from space-heating boilers if possible to maximise year-round heat demand

maximise their output. In the final column, a series of issues that might preclude the use of these technologies are provided. For the installation of any of these technologies, a feasibility study and subsequent study should be performed to understand the practicalities of installation, the potential risks and the additional costs, with the aim of producing a projected cost with a plus-or-minus 10-per-cent variable.

10.2   Photovoltaic (PV) Panels

Photovoltaic (PV) panels convert solar energy into electricity DC (direct current). As such, PV panels can either power mains electrical equipment via an alternating current inverter or can be used to recharge batteries and power DC equipment. PVs may be grid-connected or stand-alone. Stand-alone PV systems will almost always have some form of storage (batteries) and are considered to be more appropriate for use in areas of limited electrical supply.

Grid-Connected PV

For a grid-connected PV system that is designed to provide AC power to meet a proportion of a building's electrical demand, the PV cell array will need to be correctly sized to match an approved inverter. This device will connect through a circuit-breaker into a consumer unit or distribution board. Inverters should be approved for connection to the electricity-distribution network and be installed and commissioned by an accredited installer. An inverter must be able to detect the loss of power from the grid (i.e. a power cut) as there is the potential for the inverter to back-feed the grid in certain circumstances, which could cause equipment damage or risk of injury to maintenance personnel. Refer to Figure 10.1 for the system components of a grid-connected PV system.

Incorporating PV into a Building Scheme

The only accurate means of knowing how effective a PV system is likely to be is to create a theoretical energy model using real data for solar irradiance, temperature, panel-array performance and electrical-power load. This can be a time-consuming exercise and is usually carried out at detail-design phase. Initial estimates can be made on typical PV array peak power (kWp) using the manufacturer's data and average solar-irradiance data. When reviewing PV power-production estimates, it should be noted that the solar data used is usually an average over many years and that there could be a wide range in the level of performance over different years.

As previously mentioned, PV arrays produce direct-current electricity (DC) and in order to be useful within most buildings, PV arrays need to be connected to inverters, which convert to alternating current (AC). Because the inverter is not 100 per cent efficient (typically 80–90 per cent), an allowance needs to be made for the associated efficiency loss.

PV modules come in many shapes and sizes, from small, integrated roof tiles to panels that are typically two square metres in area (see Figure 10.2). There are different types of cell, constructed from silicon to differing standards of purity, providing differing levels of efficiency and cost. Crystalline silicon is the purest,

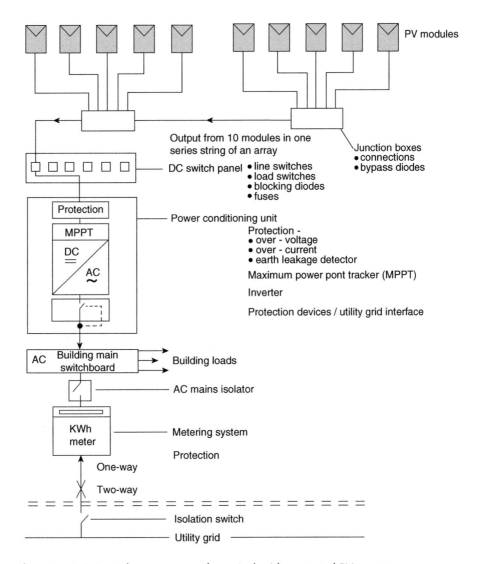

Figure 10.1: Principal components of a typical grid connected PV system.

achieving a nominal peak efficiency of 17 per cent, with polycrystalline cells achieving an efficiency of about 14 per cent. Maximum output can vary up to 250 W/m² (watts per square metre), depending on the cell technology and cell density incorporated within the module, the orientation and angle of tilt.

Within most buildings, there will be a requirement for the PV generated power to contribute to meeting the electrical demand. Larger loads would be anticipated if the building has higher energy demands, that is, air-conditioning and heavy-plant loads.

It is envisaged that PV arrays could be installed on many existing roofs, whether on accessible, flat-roof areas that would be ideal for erecting PV arrays (see Figure 10.3), or whether on pitched roofs. Structural assessments may be required on some lightweight roof constructions, however, the mounting of frames and panels

Figure 10.2: Typical PV panel roof mounted arrays.

is unlikely to constitute major, significant, additional weights. The main barrier that might need to be addressed is likely to be that of obtaining local-authority planning approval for an installation. Additional contractual issues relating to existing roofs might occur where the question of their ownership and access, particularly in respect of the contract between a landlord and tenant, make the provision of rooftop technologies more complicated.

10.3   Solar Thermal Hot-Water Systems

Active solar-heating technology involves the conversion of the sun's energy into heat by utilising an active process (e.g. passing a heat-transfer fluid, such as water, through a solar collector).

Active solar-heating technology can be used for:

- Heating water for domestic purposes;
- Space heating within buildings;
- Heating swimming pools;
- Heating water for industrial processes.

Solar space heating within buildings needs to be aligned to take advantage of the maximum available levels of solar radiation in order to provide space heating in the months in which it is most required. However, the use of solar heating for domestic, hot-water generation is a more realistic proposition and has been utilised on an increasing number of buildings in recent years; these installations

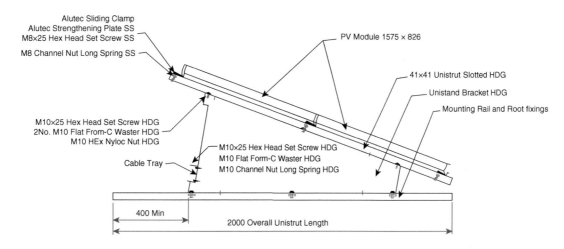

Figure 10.3: Typical PV panel mounting structure.

Figure 10.4: Roof integrated evacuated tube solar collector.

have mostly tended to be for domestic housing or for the smaller, commercial buildings.

The technology used within active solar hot-water systems is relatively mature. Recent advances in technology relate primarily to the way in which solar panels are integrated, with a move towards panels that fit into the roof structure, rather than on top of the roof tiles (see Figure 10.4).

A solar-heating system correctly sized, and during a 'typical year', should provide at least 30 per cent of the hot-water requirements for a building, with the most significant provision being between spring and autumn. The solar-heating system should be installed on the south-facing roof slope, although there are only small losses in efficiency for orientations between south-west and south-east.

Hot vapour rises to
Heat Pipe Tip

Evacuated Tube

Copper Heat Pipe

Non-toxic liquid

Cooled vapour, liquifies and
returns to bottom of pipe
to repeat cycle

Figure 10.5: Components of an evacuated tube solar collector.

Incorporating Solar-Heated Hot Water into a Building Scheme

Solar hot-water-heating systems can be installed in conjunction with conventional gas-fired boiler, radiator central-heating systems or with all-electric heating. There are three types of solar collector that are currently commercially available:

i. Selective solar flat plate
ii. Non-selective surface flat plate
iii. Evacuated tube

The evacuated tube collector (see Figure 10.5) is the most efficient, however, it is also the most expensive and the efficiency gain might not be justified by the cost increase. For the purposes of maximising carbon dioxide emission reductions, the more efficient device of the evacuated tube should be used.

Solar collectors will usually be installed on (and supported from) the slope of a roof, which should ideally be orientated south at a pitch of between 30° and 45°. Collectors can be installed on a flat roof or at ground level, but, in these situations, they will require a dedicated spatial area that should be free from shadow.

If a solar water-heating system is to be installed, a twin-coil cylinder is required, which has a solar coil at the bottom of the cylinder and a conventional coil at the top; this could take heat from the standard, domestic, hot-water circuit during the 'heating season'. Additional heat input could also be provided by an electrical immersion element. The solar cylinders are generally larger in volume and taller than conventional cylinders to facilitate stratification of the water in the tank. This is desirable as it increases the efficiency of the solar circuit.

Incorporating solar-heated hot water within an existing building might require some changes to the building, the extent of which would be dependent on ensuring

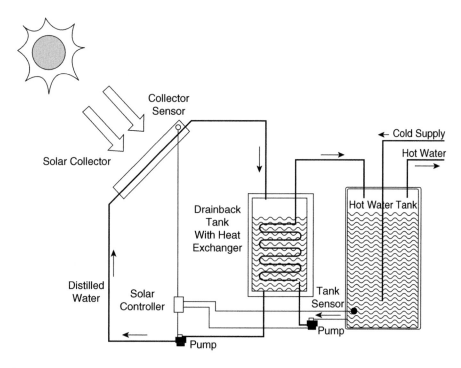

Figure 10.6: Typical solar hot water system components.

that a suitable location could be found for the solar collectors (at optimum orientation), and adapting the existing hot-water installation (usually requiring the installation of a preheat, storage vessel). Figure 10.6 shows the typical components.

The solar fraction (this is the ratio of solar-collector output against the total hot-water energy required) needs to be at least 30 per cent. In other words, the more hot water that can be provided by the solar collector, the greater the displaced, conventional-fuel energy that is saved, reducing the period over which the system pays back. However, increased solar-collector efficiency incurs increased capital cost, which, in turn, increases the payback period.

Short-period paybacks are unlikely when displacing existing hot-water systems generated by natural gas. In order to obtain payback periods of lower than 15 years, it is usually the case that solar-heated hot water must displace hot water generated by electricity. Historically, the price difference between gas and electricity has been significant; gas has been the dominant fuel for domestic and commercial hot-water heating, hence the subsidy of solar-heated hot-water installations has usually been necessary. However, with gas prices recently undergoing significant rises, this situation might change and, at the very least, the payback period is set to improve.

10.4 Wind Turbines

Wind has been harnessed to provide useful energy for many centuries. The conversion of wind energy into electricity is now a mature technology, with many thousands of wind turbines operating worldwide.

The amount of power that can be produced by a wind turbine is a function of its rated power and the wind speed at the site. The height of the rotor from the ground is also important as wind speed increases with height, and air movement is less turbulent as the height from the ground increases. A wind turbine will normally start operating when the wind speed exceeds 3 to 4 metres/second (m/s) (gentle-to-moderate breeze) and it reaches full-power output at around 10 to 12 m/s (strong breeze). At very high wind speeds (above 25 m/s), the wind turbine should shut down due to safety considerations.

Research and development, and the application of new techniques in electronics, composite materials and mechanical design, have led to more efficient, quieter and cheaper turbines.

Incorporating Wind Turbines into a Building Scheme

The potential for the installation of wind turbines onto buildings is variable, based predominantly on location and the building's surrounding activities. Small-scale turbines, typically building-mounted or with a hub height of up to 15 metres, require an uninhibited wind flow and, therefore, most urban settings are unsuitable unless they are on the coast. These turbines typically generate up to 6 kW, which is sufficient to power a large house, but is dependent on the available wind flow.

Medium-sized turbines, up to 35 m hub height, can generate up to 500 kW, sufficient to meet the needs of a small office or commercial unit. Apart from the planning issues, the turbines will require a fall distance of 1.5 times their hub height. Therefore, to enable these turbines to be erected, the site must provide approximately 50 m-diameter fall distance where there are no buildings (there is a variance of opinion as to whether car parks should also be excluded from this zone).

The wind turbine would be connected to the mains power on the customer's side of an electricity meter. When the wind is not blowing, electricity is supplied from the mains as normal. On a typical site, with an annual, mean wind speed around 5.0 m/s at hub height, the wind capacity is usually only available for about 30 per cent of the year.

Second-hand turbines are now available on the market, which have been refurbished and are provided at a much reduced cost compared to that of new turbines. The second-hand turbines are replaced on their former sites with newer, more efficient turbines, which are able to generate significantly more energy per turbine.

Planning Issues

There are particular issues relating to the siting of wind turbines, primarily based on their visual impact. Relevant information that might be typically requested by the local planning authority could include:

1. Plan(s) of the site and position of the wind turbine(s);
2. Plan of the extent of land ownership;
3. Photomontages of the turbine in the location to provide an idea of the scale and appearance so as to understand the visual impact;

4. Telecom operators require a 100-metre exclusion zone around microwave links – this will limit applicability;
5. Understanding of the ground stability.

An informal approach should initially be made to the local planning authority for an 'informal officer opinion' before proceeding to a formal application that makes the best possible case for the turbine. In planning-sensitive areas, it is usually necessary to commission a planning consultant to negotiate what can be a protracted procedure.

Site Issues

There are several constraints that can determine the location of a wind turbine. These include:

- Wind direction, obstacles and turbulence;
- Sound levels at nearby houses and buildings;
- Visual impact for local residents;
- Position of control cabinet;
- Access to site;
- Operations at the site;
- Shadow flicker;
- Location of telecommunications' links;
- Proximity of air traffic.

10.5   Biofuel Heating

Wood is a renewable and sustainable source of energy. Trees absorb carbon dioxide while they grow, thus balancing the carbon dioxide that is emitted during combustion. The carbon neutrality ensures that the biomass combustion does not contribute to global warming. The carbon dioxide that is given off during combustion is no more than would be given off if the wood were to decompose.

Energy from biomass fuels, much like that from fossil fuels, can be used to generate heat, electricity, or a combination of both. Primary fuels from biomass can be used in the form of logs, woodchip or sawdust, or processed into briquettes or wood pellets, which are suitable for combustion in boilers.

Advanced thermal conversion of biomass can produce secondary fuels such as biogas and bio-oil by gasification and pyrolysis processes. These higher-grade secondary fuels can be used efficiently to fire boilers, engines or turbines for heat and/or electricity generation.

Biomass Resources

The energy resources covered by the definition of biomass are:

Forestry/Woodland residue: round wood, branch wood and waste produced from commercial forestry management, which is derived from commercial harvesting or thinning from broadleaf or conifer forestry. Local tree surgery and local-authority waste companies may be able to fulfil fuel needs. The wood needs to be dried and chipped before it can be used within a typical combustion plant.

Energy crops: dedicated energy crops include short-rotation coppice of either the willow or poplar variety, or herbaceous grasses such as miscanthus and canary reed. They can be grown solely for the purpose of energy generation and used in the form of woodchip or secondary bio-oil/biogas fuels. Recent studies on energy crops have identified short-rotation coppice (SRC) as very suitable for biomass energy generation. SRC has the added benefit of being able to be used to help remediate land.

Issues remain over the use of biomass materials due to the poor energy transfer from the material through to the resulting energy, and also due to the take-up of land to provide the raw crops in the first instance. The latter has been blamed for contributing to the recent increases in the price of food: the argument runs that farmers are choosing to grow energy crops, thus reducing the land available for the growing of food crops. It should be recognised that there is a significant amount of natural biomass material available in addition to the availability of farmed, poor-quality agricultural land.

Wood-manufacturing residue: characterised as untreated wood waste, such as furniture-factory waste, wood-packaging waste, construction-industry waste, wood-panel waste and demolition waste. This resource can be separated into clean wood waste and wood that has been treated by preservatives or chemicals. Treated wood waste is not suitable for small-scale combustion, although it is suitable for large-scale combustion and advanced thermal processes. Clean wood waste can be used for energy in the form of wood chips, processed by extrusion into high-density wood pellets or wood briquettes or, alternatively, converted to secondary biogas or bio-oil.

Biomass Heating

Small-scale, biomass-combustion systems are categorised as being either domestic or small commercial systems up to 250 kW thermal output. Automated, small-scale, biomass-heating systems, fuelled by wood chips, offer competitive energy costs to that of electricity (and are increasingly becoming comparable to that of gas) where fuel availability and quality are consistent. Many manufacturers worldwide have developed small-scale, biomass-heating systems for domestic and commercial applications. Modern equipment, such as wood-pellet burners, has been installed successfully and provides economic competitiveness in many parts of Europe, with 100,000 modern, domestic, biomass-heating systems currently installed in Austria alone.

Biomass woodchip-combustion boilers usually have very robust stoking systems that can cope with variations in woodchip sizes, although slithers can present a problem due to bridging in the fuel-feed hopper. Maintenance of a biomass system is generally straightforward, with the ash removal taking place once a fortnight. Some boilers have a facility for quick cleaning of the heat-exchanger surfaces at the same time.

Based on a standard 30–35-per-cent moisture content, the following indicators for calorific value (potential combustion-energy content) and price have been estimated:

Wood pellets:	Wood chips (35-per-cent MC):
4.7 MWh/tonne	3 MWh/tonne
4 to 5 p/kWh	2.9 to 3.5 p/kWh

Fuel Storage

A dry storage facility for the woodchip and wood-pellet fuel is necessary; this needs to be adjacent to the boiler location so that the fuel can be fed automatically to the boiler. Combustion efficiency will be dependent on moisture content. Wood chips are typically supplied pre-dried at 20–35-per-cent moisture content. It may be possible to obtain wood chips with a higher moisture content at a lower price than dryer fuel, but it is difficult to dry out in bulk before use on a small scale; the resultant lower calorific value could make what seemed like a bargain at the time a false economy.

Initial site assessments have been based on estimates for fuel storage, using the rule of thumb of 1 kW of heat demand requiring 0.9 m³ of fuel-storage volume, including the unusable volume. Figure 10.7 shows a typical manufacturer's installation requirement for both boiler and internal fuel store.

The potential fire risk associated with storage of significant volumes of combustible material has been considered. General advice indicates that if a fuel-storage hopper is internal, the storage compartment should provide a fire rating of 90-minute resistance with fire-detection equipment and fire-extinguishing equipment being installed. In some circumstances, it might be necessary to ensure that IP ratings of electrical equipment within the fuel-hopper area meet minimum spark-protection standards. The fuel-storage room must have good ventilation for drying or reducing condensation, and for the removal of airborne particles from decomposing and fungal growths.

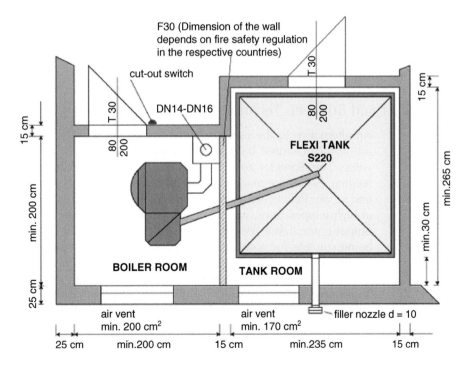

Figure 10.7: Typical small scale wood chip boiler installation with internal fuel store.

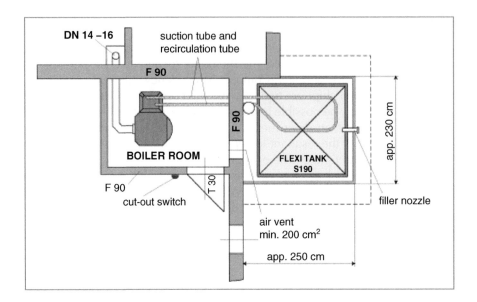

Figure 10.8: Typical small scale wood chip boiler installation with external fuel store.

An externally located fuel store is likely to offer a more practical alternative to that of internal storage. This is particularly the case when considering retro-fitting the technology to existing buildings. If the fuel storage is external, then the hopper requires shelter from rain together with good all-round ventilation.

Vandalism is a consideration in the siting and choice of fuel storage. There must also be regular cleaning of the hopper or fuel store to remove dust and decomposing deposits. In all cases, adequate access for fuel-delivery lorries needs to be available. Figure 10.8 shows a typical manufacturer's requirement for an external fuel hopper.

Wood-Pellet and Briquette Technology

An alternative to the use of wood chips is the use of wood pellets. The pellets are a manufactured bio-fuel, dried and pressed into a cylinder of high-energy value. The diameter and the length are reasonably constant to allow for easier feeding. Pellets are made out of compacted wood waste, sawdust, shavings, bark and so on. Pellets are a dry and energy-rich fuel that is burnt in small amounts at high temperatures, which means that the reaction times are fast and the heat output controllable. Wood pellets are therefore highly convenient for the user, being suitable for automated heat systems for both domestic and commercial applications (see Figure 10.9).

Pellets can be used in a wide range of stove and boiler equipment, making it a convenient choice for heating domestic properties and small buildings. Wood pellets are usually gravity-fed or screw-fed into the appliance automatically, at a rate that is varied, depending on the desired heat output from the appliance. Larger wood-pellet boiler models are usually installed within designated boiler rooms or utility spaces and are screw-fed from covered fuel stores located outside the building. As with all solid-wood fuel, the pellets must be kept dry.

Figure 10.9: Typical wood pellet fuel and delivery lorry.

Incorporating Biomass Heating into a Building Scheme

The most viable solution involving the utilisation of biomass for heating buildings may lie in the incorporation of a wood-pellet unit that operates as the lead boiler in conjunction with the existing gas-fired boilers. Such a unit could be supplied in a pre-assembled, packaged installation incorporating an integral, external fuel store or hopper. An installation such as this could then be located external to the building, making fuel deliveries more practical.

Boiler-flue locations would need to be considered carefully. Although no more onerous than typical natural-draught flues for gas boilers, flue terminals should be located so as to avoid the possibility for prevailing winds to carry smoke towards the windows of adjacent properties. This is usually avoidable if the flue terminal is installed at an appropriate height. Well maintained plant will not produce significant smoke at its working temperature, although, at start-up, pluming can be significant.

10.6   Geothermal Energy

Geothermal energy is classed as renewable on the basis of naturally occurring heat currents in the ground that dissipate to the environment. There are, however, many geothermal systems that are, strictly speaking, not renewable since their heat currents are increased artificially (e.g. by drilling into hot aquifers); in this case, the supply of heat energy is not renewable at the full extraction rate on a long-term basis and it is finite in its availability.

Groundwater occurs naturally as part of the water cycle and as it flows through the ground towards rivers and seas, its temperature tends to match that of the ground. The water is used to extract heat energy from the ground. Groundwater flows through aquifers, which typically consist of gravel, sand, sandstone or fractured rock. These are permeable because they are porous and have connected spaces that allow water to flow through. This water can be extracted through a well, drilled into the aquifer.

The position of an aquifer varies from one location to another. To establish whether groundwater is available, a hydrogeological map is used, which shows how productive an aquifer will be. However, to determine its productivity, more details of its properties are required, such as transmissivity and hydraulic conductivity; in addition, the well's flow rate will vary depending on the type and size of the well – a good flow rate is 25–50 l/s. Test boreholes would need to be carried out at any site to confirm production rates before commitment to a groundwater scheme could be made.

Responsibility for the management of groundwater resources lies typically with a government agency, and any groundwater scheme needs to comply with its rules. An abstraction licence may also be necessary together with an environmental-impact assessment and, in some cases, water must be re-injected back into the aquifer.

Applying Geothermal to Building Schemes

The ground offers a good resource for the heating and cooling of buildings. From a depth of 4–200 m, the ground temperature is relatively constant in temperate zones at around 12°C irrespective of the season.

Ground, coupled heating and cooling systems can be either:

- **Open-loop systems**: where water within the ground is used directly to provide cooling, or
- **Closed-loop systems**: where heat exchangers are constructed within the ground to extract energy to be used in a heat pump to provide cooling or heating.

Open-loop cooling systems require minimal energy input compared with traditional refrigeration systems. They have a reported coefficient of system performance of between 100 and 10, depending on the application, compared with around 3 for a traditional system. As a result, open-loop systems compare favourably with conventional cooling systems in terms of running cost, energy consumed and carbon dioxide emissions.

In situations where the aquifer is unproductive in terms of groundwater production, or where the water quality is unsuitable for cooling, or the extraction costs too high, closed-loop systems may be considered, whereby the heat exchanger is constructed within the ground.

Closed-loop heat exchangers usually consist of a sealed loop of polyurethane pipe containing a heat-transfer fluid that is pumped around the loop. The heat exchanger may be installed vertically or horizontally, but the choice of the arrangement will depend on the available land, local soil type and excavation costs.

In order to determine the length of heat exchanger needed to meet a given load, the thermal properties of the ground need to be assessed. A geotechnical survey should be carried out to reduce the uncertainty associated with the ground's thermal properties. For large schemes where multiple boreholes are used, a trial borehole and properties' test may also be appropriate.

10.7   Combined Heat and Power (CHP)

Combined heat and power (CHP) draws together two processes, namely, the generation of electrical power using a **prime mover** (vehicle used to power the electrical-generation process – such as a combustion engine or turbine) and the production of hot water or stream from heat recovered from the prime mover. The optimisation of electrical-power generation, in conjunction with useful heat recovered and the amount of fuel used to drive the process, has traditionally formed the basis of assessing CHP efficiency.

The overall CHP installation efficiency should also take into account the level to which generated heat is used without waste and the amount of electricity that can be used on site (without recourse to export to the national grid). An overall judgement on CHP efficiency might also include the environmental emissions (direct and indirect) associated with the fuel used by the prime mover that displaces the fuel that would have been used in conventional, separately generated heat and power. Factors other than those directly related to thermodynamic efficiency, such as economic or fiscal issues that dictate the level of investment required to achieve the highest operational efficiency, might also be considered.

Unless the fuel used by the CHP prime mover is from a renewable source (i.e. a biofuel), the CHP system would not be classed as a renewable technology. However, CHP (albeit systems that might not utilise a renewable fuel) has the potential, in the right circumstances, to offer overall reductions in carbon dioxide emissions, due to the value of the heat produced and utilised within a building, rather than its being wasted to the atmosphere. It is this ability that enables CHP to act as a carbon-mitigation technology and enables it to be considered as part of a programme aimed at minimising carbon emissions.

Incorporating CHP into Buildings

CHP has traditionally been applied on a comparatively large-scale-generation capacity to the extent that even a modest installation would have usually only been considered for sites of a campus nature, such as a university or hospital, or for sites with major year-round heat demands, such as leisure-centre or swimming-pool complexes. More recent developments in prime-mover technology have enabled CHP to be scaled down to a size that can more readily be applied to a wider range of buildings.

CHP, typically capable of providing heat outputs of between 5 and 20 kW (and electrical outputs of between 2 and 40 kW) is now commercially available and in use in many buildings that would otherwise not be considered suitable for a CHP application.

CHP units would be installed adjacent to the existing boiler location and connected into the existing heating system so that maximum benefit from recovered heat during the units' operation could be obtained. In practical terms, sufficient space for the CHP unit(s) would need to be found adjacent to, or within the near vicinity of, the boiler room; consideration should be given to the installation of the CHP flue (allowing the products of combustion to be expelled), together with ensuring that sufficient capacity in the building's gas supply exists to fuel the CHP unit. Figure 10.10 shows a simplified component-and-connection arrangement.

Optimising CHP Plant Operation

Conventional heating boilers, electrical generators or power-supply-capacity provisions are normally sized so that they will cater for the worst-case, maximum

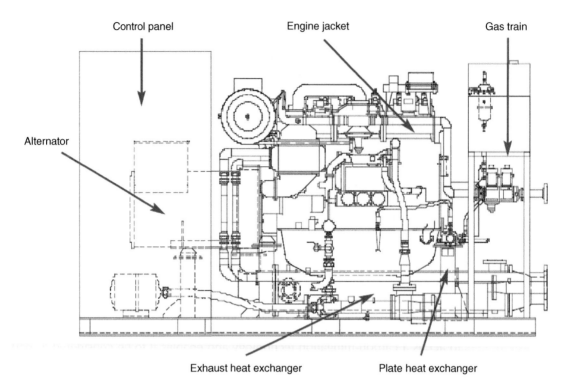

Figure 10.10: Typical arrangement for CHP.

heating or power demand. The plant will then cover all eventualities by modulating output, or turning on and off, to suit any demand that is less than the maximum peak capacity. Conversely, CHP plant capacity is usually sized to meet the minimum, or base, heating or power demand. This is because a building's peak demand for heat and power usually occurs only for a comparatively brief period and is influenced by the time of day, building-use pattern and seasonal climate. It is more difficult to modulate CHP output to match either heat or power demand, since these demands might not always coincide with one another.

Over a year, a building's typical heat-demand profile will peak during the winter months and fall off during the summer, when the heat demand is limited to that which is required to satisfy domestic, hot-ater consumption (assuming that this is serviced indirectly from the central-heating boilers). Although wintertime demands the increased use of power for lighting, summer might precipitate an increase in power consumption associated with ventilation or air-conditioning systems.

A CHP system, sized for peak electrical-power demand, would enable any requirement for importing electrical power from the national grid to be mitigated. A high generation capacity enables a proportionally high heat-generation capacity (within the limits of the prime mover power-to-heat ratios). However, during periods of limited electrical demand, the system might not be economical to run unless export of electricity to the national grid is possible. A correspondingly low heat demand might require the unit to be switched off, if the heat cannot be utilised for another process and is simply rejected as waste heat to the atmosphere.

The alternative is to select a CHP plant to operate on base-load power and heat demand. This enables the maximum, potential CHP operating hours to be achieved. However, for periods of peak electrical and power demands, imported grid electricity and additional heating will be necessary.

The energy ratio is the amount of heat simultaneously produced in relation to the electrical power generated. Typically, CHP heat-to-power ratio implies low electrical-generator efficiency, as less fuel energy is converted to work, with the converse being true for low heat-to-power ratio. This has led to some CHP systems to be oversized in order to capture the economies of scale and make plant more economical, tending to move CHP away from the ideal heat-to-power ratio that achieves optimum efficiency.

By using small-scale CHP units, sized to ensure that the units can operate for the longest period possible, it is intended to match CHP capacity to heat loads in the most efficient manner possible.

The CHP technology is based on conventional, well proven technology that uses natural gas as fuel for the prime mover. Whilst this is recognised as a reliable form of fuel, it must be remembered that natural gas prices are on the rise and, therefore, commercial viability is likely to be affected in the short-to-medium term.

10.8   Heat Pumps

A heat pump is a device for transferring heat from a lower-temperature heat source to a higher-temperature heat sink. This is the opposite of the natural flow of heat from a hot source to a cold sink, but is made possible by the application of an external energy source to drive a thermodynamic refrigeration cycle. The important characteristic of a heat pump is that the amount of heat that can be transferred is greater than the energy needed to drive the cycle.

This is the same process as that in a fridge or in an air-conditioning unit. In the case of a fridge, the heat energy is pumped from the interior of the fridge to the elements at the back. Removing this heat energy makes the interior of the fridge cold and the elements at the back warm. As the elements become warmer than room temperature, the heat energy (which was originally inside the fridge) is lost into the air of the room. A heat-pump heating system does exactly the same thing, though on a bigger scale, and removes its heat from a source outside the room – such as the outside air, or the ground.

The ratio between the heat provided to the sink and the energy required is known as the coefficient of performance (COP). Electrically-driven heat pumps used for space-heating applications in moderate climates usually have a COP of a least 3.5 at design conditions. This means that 3.5 kWh of heat is the output for 1 kWh of electricity used to drive the process. The COP is the determinant of whether the heat pump will be more economical to use than an alternative heating appliance and whether the carbon emissions will be less than those from an alternative heating appliance.

In simple terms, such a heat pump will be cheaper to operate, provided that the electricity price is no more than 3.5 times the price of an alternative fuel. There are other factors to be considered in a more detailed analysis of the benefits, such as maintenance costs and equipment life, but the fuel–price ratio is the key. This is also the main reason why the heat-pump market in some countries has developed much more quickly than others where, historically, electricity has been more than 3.5 times the cost of natural gas. However, the long-term trend is for gas prices to increase faster than electricity prices, whilst heat-pump COP gradually improves to provide a clear operating-cost advantage for heat pumps. Heat pumps are already cheaper to operate than are oil and LPG, and much cheaper to operate than is direct electric heating. Whilst the operating costs for heat pumps and condensing boilers are rather similar at current fuel prices, the case for heat pumps as a low-carbon technology is more conclusive.

The heat pump emits 35 per cent less carbon dioxide than the condensing boiler and can therefore be considered appropriate for inclusion.

There are three main types of heat pump, where the system takes heat either from the air, the ground, or from water (from ponds, rivers and boreholes).

Ground-Source Heat Pumps (GSHPs)

The two most common types of GSHP systems are known as horizontal (or trenched) ground loops and vertical borehole loops. An electrically-driven pump circulates a mixture of water and glycol through the coil, extracting heat from the ground. The size of the ground loops, whether horizontal or vertical, needs to be matched to both the peak heat demand and annual energy requirement of the property. The longer the loop, the more energy the pump is capable of producing. Vertical boreholes work on the same principle, but the plastic tube is arranged in a U-shape going downwards into the ground. Boreholes can be anything from 15 to 100 metres deep, depending on the heat demand from the building.

Ground-source heat pumps, once installed, are unobtrusive. As the ground maintains a more constant temperature than does the air, GSHPs are not prone, like ASHPs, to fluctuations in efficiency in colder weather. However, installing the ground loop can be expensive, and its sizing depends on geological factors – a real issue in the UK, where local geology varies widely across locations.

Air-Source Heat Pumps (ASHPs)

Most ASHPs are sited just outside the property. An electrically-driven fan draws air across the evaporator, cooling the air stream and supplying heat to the heat pump. Below about 7°C, ice can form on the evaporator as the air is cooled, restricting the airflow and impairing performance. For this reason, ASHPs always include a defrosting cycle. A common defrosting method is to extract heat from the heat sink (the building or hot-water tank) and resupply the heat to the evaporator to melt the ice – in effect, operating the heat pump in reverse, so that the evaporator becomes the condenser, and the condenser becomes the evaporator. While this is happening, not only is heat being taken from the building, but also no heat is being sent to the building, which might temporarily lower the heat pump's COP.

Another potential heat source is air from the building, extracted by an exhaust-air heat pump. Exhaust-air systems have the advantage that their heat source has a fairly constant temperature of around 20°C, but they need to be carefully designed to maintain a balance with the ventilation requirements of the building. They are usually suitable only for well insulated buildings.

Endnote

1 Jacob Funk Kirkegaard, Thilo Hanemann, Lutz Weischer, Matt Miller; Toward a Sunny Future? Global Integration in the Solar PV Industry; World Resources Institute, May 2010

11 Embodied Carbon

There is a global desire to reduce the energy consumption within operational buildings through a variety of legislation, incentives and use of renewable technologies. Many countries are looking to implement zero-carbon buildings from an operational perspective. However, few of these standards take account of the carbon resulting from the materials used in the construction and refurbishment programmes.

The study of embodied carbon looks at the inherent carbon contained within the materials used on site, from their initial extraction through a variety of processes to their final use on site. Whilst a number of systems and standards are in place, there is little compatibility between the standards, making comparison difficult and an accurate assessment challenging.

However, with the current focus on energy reduction, a recognised industry standard is being developed by the EU with a means to enable not only a standardised approach, but also to choose materials based on their embodied-carbon value.

Chapter Learning Guide

This chapter will describe embodied carbon and some of the challenges driving the need for a standardised approach.

- Definition of embodied carbon and some of the issues surrounding the extent and limit of its scope;

Sustainable Refurbishment, First Edition. Sunil Shah.
© 2012 Sunil Shah. Published 2012 by Blackwell Publishing Ltd.

- Impact of zero-carbon regulations on the embodied carbon within buildings;
- Overview of CEN/TC 350 which will include an embodied-carbon standard;
- Approaches to reduce embodied carbon within buildings.

Key messages include:

- Embodied carbon forms the significant part of low-operational assets such as infrastructure and stadiums;
- Voluntary means, and latterly, legislations, will be in place to reduce the embodied carbon of buildings;
- To properly assess the true embodied-carbon impact of a material, a life-cycle view must be taken.

11.1   Introduction

With the current focus on the operational use of energy during the use of the building, largely calculated at the design stage, legislation is in place to reduce this significantly. The UK has a target for zero-carbon housing by 2016, with many European countries and some US states looking at similar targets. Non-domestic, zero-carbon targets are not far behind, with the UK proposing a 2019 date. As operational energy reduces, the energy required to refurbish the buildings, including that for the additional materials, will gain greater relevance.

Energy is required for the extraction of the raw materials, processing, transportation and fabrication of the construction materials and equipment. This is the embodied energy (see Figure 11.1) and can be defined as:

Embodied energy (carbon) is defined as the total energy consumed (carbon released) from direct and indirect processes associated with a product or service through its building lifecycle from construction, maintenance, refurbishment, demolition and safe disposal. The boundaries are otherwise known as cradle to grave. This includes all activities from mineral extraction (quarrying/mining), manufacturing, transportation, fabrication processes, until the product is ready to be used on site. It will also cover the maintenance requirements over the project lifetime necessary to maintain the product together with its demolition and subsequent safe disposal[1].

Such an approach will enable the whole life-cycle analysis of carbon to be made for a building, capturing the embodied and operational energy.

However, such a definition does not take into account the reuse and recycling of materials, arising from demolition, that will be particularly relevant during refurbishment projects. Such materials will have a lower embodied-carbon value as they are reworked for further use. Utilising an arbitrary carbon-reduction factor is one way in which to take this into account. Such an argument has been used by the steel industry, where a majority of the steel in use will later be recycled for reuse.

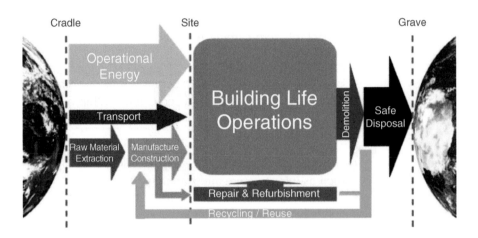

Figure 11.1: Lifecycle approach to carbon footprinting of materials.

Defining 'embodied carbon' accurately is complicated. For example, a concrete block sourced from a factory just down the road will contain less embodied carbon than one imported from China because of the difference in the energy used to transport it. An aluminium window might contain relatively little embodied carbon if it is made from recycled materials, or from virgin aluminium produced by hydroelectricity, rather than by carbon-intensive coal. The term, 'embodied carbon', also includes emissions from the contractors' operations, ranging from those from their excavators and cranes to those from boiling the kettle for all those cups of tea.

As energy-efficiency legislation and building regulations reduce the operational energy demand further towards a zero-carbon goal, the level of embodied energy within a building will increase substantially. However, the level of carbon within the materials through their life cycle will vary significantly between varying building types (see Figure 11.2a and 11.2b). For infrastructure schemes where there are limited operational activities, the impact of zero-carbon developments has a minimal effect. Where buildings have a greater level of operational activity, such as in housing and offices, the potential impact from the zero-carbon agenda will increase significantly and, therefore, the embodied carbon becomes a greater issue.

This presents an opportunity, in the case of developments with a majority of infrastructure and low-operational activities, to measure their embodied carbon with a view to replacing materials with those with a lower carbon content.

11.2 Embodied Carbon Standards

Much work has been done across the industry to quantify the embodied carbon in construction. However, there is no commonly recognised method of simultaneously analysing both the embodied and operational carbon emissions with respect to any single time period. This raises a problem in that there is no consensus of opinion on embodied-carbon valuations due to the vested interests of many of the providers.

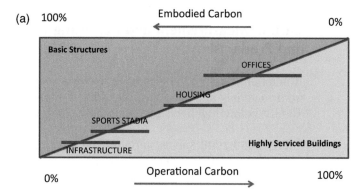

Figure 11.2a: Relationship between embodied and operational carbon for typical buildings.

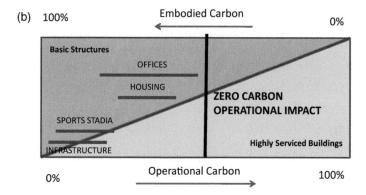

Figure 11.2b: Impact of zero operational carbon on typical buildings carbon footprint.

All the available tools calculate the carbon footprint of buildings, but rely on third-party data for the embodied carbon of materials. Bath University has published a database for the embodied-carbon content of construction materials, which is used by many organisations. Unfortunately, this data is generic and not product-specific. As carbon profiling becomes more sophisticated, the ultimate goal should be the availability of robust data that allows specifiers to choose, for example, an aluminium window, rather than an equivalent alternative product, on the basis of embodied-carbon data. This requires the agreement of product manufacturers to accept a single standardised way of measuring embodied-carbon content.

New environmental and carbon-footprinting standards from recognised bodies, including the International Standards Organisation (ISO), World Resources Institute (WRI) and the World Business Council for Sustainable Development (WBCSD), are helping to provide a better definition. Of most relevance is the CEN/TC 350 series, developed across Europe in response to the Construction Directive.

Carbon Measurement and Reporting of Materials and Products

A number of methodologies have been developed to measure the carbon from materials and products, including a number of proprietary systems. The list below highlights the commonly cited and accepted standards:

- **PAS 2050:2008** *Specification for the assessment of the life cycle greenhouse gas emissions of goods and services* is a general carbon-footprinting standard, not specific to buildings, but clearly written, capturing emission factors, and widely used. A similar standard is well developed in France.
- **BS EN ISO 14021** *Environmental labels and declarations – Self declared environmental claims (Type II environmental labelling)*
- **BS EN ISO 14044:2006** *Environmental management –Life cycle assessment – Requirements and guidelines,* Clause 4.3.4.3
- **BS EN ISO/IEC 17050-1** *Conformity assessment – Supplier's declaration of conformity – Part 1: General requirements*
- **ISO/TS 14048:2002** *Environmental management – Life cycle assessment – Data documentation format,* Clause 5.2.2
- **IPCC 2006** *Guidelines for National Greenhouse Gas Inventories*

CEN/TC 350

The lack of a consistent approach has led to a number of issues, which the CEN/TC 350 approach is looking to resolve. In particular, these issues are:

- Consistent product data – what are things actually made of?
- Consistent carbon data – what is the impact?
  - Manufacturing processes, energy sources, location/transport etc.
  - Bio-renewable (organic) materials;
- Scope of assessment – including accounting for recycled content and recyclability;
- Addressing shortcomings in 'global' indicators for buildings – for example, quantity (kg) of whole-life carbon dioxide per square metre of floor area.

CEN/TC 350 was created in 2005 in response to the standardisation mandate (M/350, 29 March 2004) from DG ENTR of the European Commission, directing CEN to 'provide a method for the voluntary delivery of environmental information that supports the construction of sustainable works including new and existing buildings'.

The aim of the standard is to provide a harmonised, horizontal (that is, applicable to all products and building types) approach to the measurement of embodied and operational environmental impacts of construction products and whole buildings across their entire life cycle. Included within the calculation will be the economic performance and social performance of buildings with reporting on many sustainability parameters.

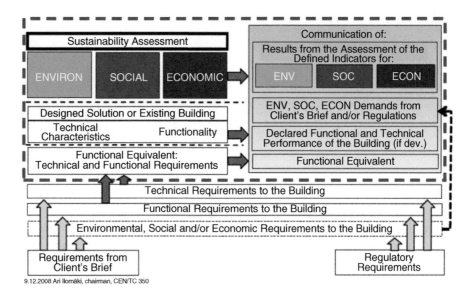

9.12.2008 Ari Ilomäki, chairman, CEN/TC 350

Figure 11.3: Concept of CEN/TC 350.

It is anticipated that the CEN/TC 350 rules for product declarations will come into force in 2013. The concept design of the standard, as described in Figure 11.3, seeks to pull together the environmental, social and economic parameters into a reportable and standardised approach. Figure 11.4 illustrates the boundaries associated with the system, where the four stages represent the life cycle of the building with an additional module for the further use of the materials, which may influence the final carbon footprint of the materials.

It should be noted that, within the use phase, there are specific sections on repair, replacement and refurbishment, which mirrors the four-stage, sustainable-refurbishment approach taken. It therefore provides the ability to define the carbon footprint of specific building-refurbishment options to enable informed decisions to be made.

Individual practices are converting the raw data to provide kg-CO_2e figures for resources, components, measured units and measured composites. This will allow the embodied carbon to be estimated from the same measurement and procedures as those for costs. Over time, this will produce benchmark data at the building level, coupled with additional information as European legislation will require data to be provided by product manufacturers by the mid-2010s.

To add approximate cradle-to-gate CO_2e values to a price database, you must have the following information. The calculations are simple, if you have the data, but there is a huge amount of data to collect.

• Material compositions of priced products and their packaging;
• Weights of constituent parts (and what material-density and fuel-mix values were applied in deriving CO_2 and CO_2e values from energy values);
• Average percentages of recycled materials.

Stage 1: Product			Stage 2: Construction		Stage 3: Use							Stage 4: End of Life				Benefits and Loads Beyond the System
Raw Material Supply	Transportation of Materials	Manufacturing	Transportation of Product	Construction/ Installation	Use of Products	Maintenance	Repair	Replacement	Refurbishment	Operational Energy Use	Water Use	Deconstruction/ Demolition	Transportation of Waste	Waste Processing	Waste Disposal	Reuse, Recycling, Recovery Potential

Figure 11.4: TC350 boundaries.

Beyond this level, the calculations become very site-specific, relating to the storage, use and operational activities at the construction site. Information can be obtained using the BREEAM or LEED methodology to determine site-construction impacts.

A number of proprietary tools are being developed by commercial providers and also in-house by major asset managers. These will certainly need to align with the CEN/TC 350 approach across Europe. The idea behind these tools is to enable users to adjust a single building element and for the tool to automatically calculate the impacts elsewhere. This will enable comparison of building materials and the achievement of calculated carbon reduction from materials' embodied carbon, affecting life-cycle and end-of-life decisions.

11.3 Varying Embodied Carbon Values in Buildings

Evaluating emissions requires an understanding of how the period in time at which they are generated relates to the time when the benefit is derived. Turning on a light bulb is quite easy to conceptualise – the emissions are being generated 'live' in the moment of use by the user's turning on of the switch. However, the emissions generated by manufacturing a carpet might have occurred perhaps months before its use and this expenditure might also then be shared over the seven (or so) years of its life. Another point to note regarding the impact of time is that the built environment around us represents an existing resource of spent carbon. Indeed, part of the fact that global warming is currently happening is down to its existence. With this in mind, decisions to replace materials or equipment should be measured against the benefits of any new proposals and the length of time that they, in turn, may last.

Depending on the building type, different buildings can vary significantly in their embodied-carbon content. Distribution warehouses do not use much energy for heating and lighting, which means the embodied-carbon component of the building's total carbon footprint is high. The embodied carbon in a distribution warehouse has been calculated as 60 per cent of its total lifetime carbon footprint, whereas a supermarket, which uses lots of energy, has an embodied-carbon content of 20 per cent; a house has a content of something like 30 per cent (see Figure 11.5).[2]

Finally, understanding the lifespan of building components or systems is crucial to the understanding of their embodied-carbon efficiency. The key issue is the

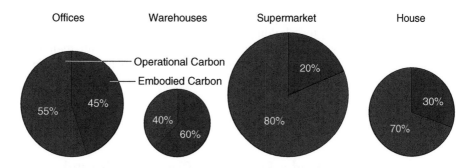

Figure 11.5: Typical split between operational and embodied carbon emissions.

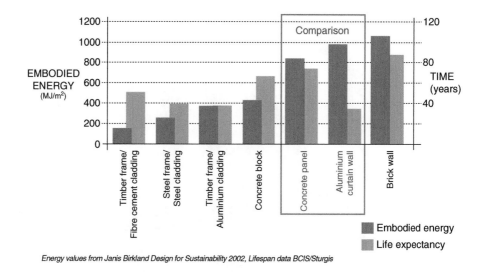

Energy values from Janis Birkland Design for Sustainability 2002, Lifespan data BCIS/Sturgis

Figure 11.6: Lifespan and embodied energy of different cladding systems.

overall carbon impact of the material during its lifetime. Hence, a high-embodied-energy product, such as a brick, might have a lower overall impact than does timber cladding, because it lasts hundreds of years. Only by factoring in their anticipated lifespan, can the overall carbon footprint, or embodied-carbon efficiency, of materials be calculated.

For example, in the comparison of concrete and aluminium cladding systems[3], both had similar embodied-energy investments (see Figure 11.6), but their anticipated lifespans were very different. When examined using BCIS[4] data, the concrete system had roughly twice the life of the aluminium system. In this example, it could therefore be said that the concrete system had twice the embodied-energy efficiency of the aluminium. This sort of information is vital to the designer who is selecting components for their durability, and who needs to be able to carry out a comparative carbon analysis of their relative efficiency.

A study examined recurring embodied energy and found that by the time a typical office building is 50 years' old, 144 per cent of the initial embodied energy will have been spent again through maintenance and replacement of fabric[5]. Crucially, the cladding finishes and services were identified as the biggest components of these recurring carbon emissions, with their component lifespans being the biggest coefficient in determining the magnitude of these recurring emissions.

Case Study: Ropemaker Place, British Land

Figure 11.7: Ropemaker Place.

Ropemaker Place is an 81,000 m², £155-million, low-energy, commercial-office building in the City of London. British Land prided itself on the low carbon emissions of its new development but, like most of the industry, had little idea of how much carbon was embodied in such buildings. Based on Part L calculations for energy use, it was found that the building's embodied carbon represented 42 per cent of emissions, and its operational carbon represented 58 per cent, based on a sixty-year lifetime. As the UK government looks to decarbonise the grid electricity, the embodied carbon could increase to as much as 70 per cent over the building's lifetime.

Assessed against British Land's total emissions for the last two years, in the first year, 2008–09, the embodied carbon from its new developments was the same as all its operational emissions, which includes energy used by tenants. In the second year, 2009–10, a quieter development year, embodied-carbon emissions were still at 50 per cent of its operational emissions.

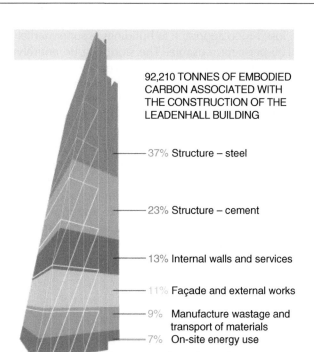

92,210 TONNES OF EMBODIED
CARBON ASSOCIATED WITH
THE CONSTRUCTION OF THE
LEADENHALL BUILDING

37% Structure – steel

23% Structure – cement

13% Internal walls and services

11% Façade and external works

9% Manufacture wastage and
transport of materials

7% On-site energy use

British Land are intending to assess the performance of materials more carefully in the future. The first stage is by more thorough monitoring – for example, by determining the percentage of cement replacement used at Ropemaker Place. In future, British Land will expect the structural engineer and contractor to work out the optimum amount of cement replacement that can be used and determine the feasibility of reusing materials from buildings that are being demolished. Other changes will include making sure that all the components in a system have a similar lifespan. For example, the life expectancy of the cladding system can be dictated by that of its fixings, which might fail before the rest of the system. Roofing materials and sealants have comparatively short lives.

In a study for Marks & Spencer[6], strategies were developed to deliver a zero-embodied-carbon solution that draws on innovative sustainability concepts such as industrial ecology, cradle-to-cradle design and bio mimicry to minimise, and eventually negate, impacts that are harmful to the environment. Closed-loop systems are a conceptual approach to managing the entire life cycle of a product or building, whereby all materials not safely consumed in the use of the product are designed to be reused at the end of their life. As a result, waste elements that cannot be eliminated are either recaptured and reused in the process of making the same or other products, or made biodegradable to input into the broader biosphere.

To reduce the environmental burden, closed-loop systems have the potential to reduce capital costs through material and energy savings. By adapting such a closed-loop approach to buildings, a supermarket store could potentially become 100-per-cent reusable. The store can be entirely demountable at the end of its life, which is typically after 60 years. All the store's components could be reused within its supply chain, either for new buildings or to maintain existing buildings with little or no reprocessing. Although the carbon 'hit' will need to be absorbed within the first life cycle of the building, subsequent life cycles will emit only a limited amount of carbon.

This closed-loop approach would not be feasible without a wholesale rethinking of how we design, procure and manage buildings throughout their full life cycle.

As the economic and environmental benefits of buildings that use resources more efficiently become better understood, the need for immediate action will be recognised. We will see life-cycle optimisation of buildings becoming the norm. However, this will take a combined effort from a collaboration between private developers and occupiers as well as between national and regional governments.

CASE STUDY 1

London Fire and Emergency Planning Authority (LFEPA): Energy Case Study

Reproduced by Permission of the London Fire Brigade

Carbon Reduction at LFEPA

With 113 fire stations across London, a brand new headquarters building in Southwark housing 800 staff and a fleet of over 600 emergency vehicles, London Fire Brigade (LFB) has a significant impact upon the community and the environment. An ambitious programme to be greener began in 2003 and by 2010 the programme has seen carbon emissions drop by 21% from 1990 levels. This is despite a significant year-on-year growth in both staff numbers and functions undertaken within the fire brigade.

Being a fire and rescue service, the brigade's core aim will always be to safeguard life, whether saving people from burning buildings or cutting them free from road accidents, but winning the prestigious overall award at City of London Corporation's Sustainable City Awards in 2007 showed that LFB also has sustainability at the heart of its agenda. LFB's buildings and vehicles are now much greener, all of the furniture from the old headquarters was reused or recycled, the take-up of a Bikes 4 Work scheme amongst staff has been impressive and the brigade's commitment to tackling climate change is rubbing off on its suppliers.

The brigade achieved the Carbon Trust Standard in 2008 in recognition of its work in lowering carbon emissions and tackling climate change, and was identified as an 'exemplar' organisation in its drive to reduce London's carbon footprint.

A winner of various industry awards, LFB is now a symbol of best practice in the public sector.

Energy Manager Ian Shaw said: 'We recognised the need for our green programme at a very early stage and haven't stopped since. We have found endless ways of bringing down carbon emissions, some innovative, some more traditional, but all have been effective and our staff have really taken up the green mantle.'

The brigade has been fitting out fire stations with on-site energy-generation technology such as photovoltaic (PV) systems, wind turbines, solar thermal, and combined heat and power. To date, 27 photovoltaic, 14 solar thermal, 2 wind turbines, 36 CHP schemes, and 82 high-efficiency lighting installations are now in place at London fire stations. Improvements are also being made to heating systems, and sustainable wool insulation is being used in lofts. Currently, around 45 of the 112 buildings are fitted with sustainable-energy sources providing around 6% of the electricity that the brigade uses.

The programme has also demonstrated that renewable and low-carbon technologies can be successfully applied to existing and, in some cases, ageing buildings. Tooting Fire Station, for example, was opened in 1907 and is Grade II listed.

Added to this, a staff-motivation campaign – entitled LFB Green – has been designed to further reduce energy and water usage. We have recruited and trained over 220 'green champions' across the brigade who are responsible for ensuring that their colleagues are on-board in terms of green issues. They are responsible for checking that sites, or fire stations, are as environmentally friendly as possible. They audit their sites and report back with suggestions as to how things can be improved. They are supported by the 'Green Zone' on our intranet site, providing details of recycling schemes, energy-saving tips and transport initiatives.

In rolling out its sustainable-procurement policy, LFB has increased year-on-year its spend on goods with recycled content. Hundreds of products on the brigade's internal ordering system have changed over to 'green/recycled content', making staff more aware of environmentally friendly products and doubling the purchase rate of these products. As a result, there has been an increased participation in in-house recycling initiatives, backed up still further by a new centralised recycling contract for all 113 fire stations. Figures for the past year show that the brigade has recycled over 49 tonnes of paper, 1800 kg of batteries and 1200 kg of clothing, not to mention old fire hoses being used to make belts and other items, rather than being sent to landfill.

By the brigade's efforts in minimising the organisation's effects on climate change, communicating its environment policy amongst partners, major contractors and suppliers, and advising them of internal policies and approaches to the environmental-management system, LFB is encouraging a wide audience to follow its best practice in this area and improve environmental performance.

How Did We Go About Reducing the Brigade's Carbon Emissions?

- We have invested £4.4 million into a revolving 'energy efficiency' fund, which means that when we make financial savings by saving energy, the money is then reinvested into further energy-saving initiatives. A further £500,000 of grants has also been secured to carry out further energy-efficiency work
- We are introducing energy-saving lights into all of our buildings, with many already upgraded, as well as light sensors so that the lights automatically switch off when there's no one in the room. This has drastically reduced our carbon footprint.
- We are working towards making all of our buildings and fire stations as energy efficient as possible. This has been a challenge as many of our fire stations are old, listed buildings. Of our 113 stations, 99 are now rated typical, or good, against the CIBSE benchmarks for energy efficiency. Our average DEC rating is D 85 with 4 achieving a B rating.
- We have carried out a programme of improving existing buildings, for example, window replacement, replacing inefficient lighting, installing sensor lighting, replacing old, inefficient heating systems, and adding extra loft, wall and pipe insulation.
- We are updating and improving our heating systems, fitting energy-efficient combined-heat-and-power (CHP) systems and new thermostats whereever possible.
- 'Grey' water recycling for toilet flushing is being introduced into stations when they undergo major station refurbishment, and other water-saving measures are being installed, such as manual-push-to-flush urinal controls.
- We have developed a station design brief to aim for all new buildings to be built to the BREEAM excellent environmental rating.
- We introduced a new recycling scheme into all our stations with clearly marked bins so that over 50% of waste is recycled. We also send a lot of our used equipment away to be recycled, for example, we sent some 250 tonnes of old furniture to be reused when we moved into our new headquarters in early 2008, and old fire hoses are sent to a company that uses them to make handbags and belts.
- Over 2000 staff have participated in the Bikes 4 Work scheme. Staff can get tax-free bikes, thus encouraging them to cycle to work, rather than drive.
- In being more energy efficient, we have made savings of £165,000 for 2009–10. We have seen a CO_2 reduction of 21% since 1990 (our original Greater London Authority target was a reduction of 15% by 2010). We aim to reduce our carbon emissions by 60% by 2025.
- LFB have installed photovoltaic (PV) panels at several fire stations. The first PV panel was installed at Richmond Fire Station in 2005, and now generates around three-quarters of the station's electricity – PV panels are now at 27 stations.

- We installed 2 x 45 kW pellet biomass boilers with a 600 l thermal store to provide all the heating and hot water for the station and offices at Croydon Fire Station as part of its refurbishment. Pellets are from a sustainable source and the large pellet store reduces the number of deliveries required, expected to save over 70 tonnes of CO_2 per annum (cost neutral).
- The installation of lighting controls has shown an average saving of 20% (£3,000 per annum per site and payback of 4.5 years), together with high-efficiency light fittings showing an average saving of 8% (£1,200 per annum per site and payback of 3.2 years).
- Loft, cavity-wall and draught-proofing insulation have contributed to savings of up to 25% in gas usage (saving £10,100 per annum and payback of 4.6 years).

Installation of CHP

Between April 2007 and September 2008, 19 London fire stations had mini-CHP units installed as part of a package of energy-efficiency and LZC installations at a cost of £726,113. The relatively high water consumption and the 24-7 operation of fire stations make then ideal sites in which to install small-scale CHP. As at 2011, 36 units had been installed at 33 stations.

The initial installations have proved successful and are being used as a template for future projects. Typical example: between July 2007 and August 2008, the Battersea Fire Station's mini-CHP unit has generated 27,018 kW of electricity at an efficiency of 78%. If this figure is repeated across all 36 fire stations, this would represent a saving of 520 tonnes of carbon per annum. Based on average consumption over the previous four years, the site's overall energy consumption has reduced by 8.91%, CO_2 emissions by 23%, and energy costs by 26%.

Each system consists of a Dachs mini-CHP unit, electricity rated at 5.5 kWe and thermal rating up to 12.5 kWth, a condenser and a buffer vessel. The benefit of installing the condenser is that the thermal output is increased and less flue gases are produced. The 750 l buffer vessel not only stores unused energy, but also creates a heat demand that keeps the Dachs unit running. Each system is installed so that it is connected to both the low-loss header on the heating circuit and the secondary return on the hot water, thus maximising year-round running. The controls are set so that the CHP always acts as the lead boiler for both heating and hot water.

Owing to the standardised design, the installation is now being rolled out to more stations very quickly, with only such items as the pipe runs differing between sites. This allows for some off-site fabrication and rapid installation with minimal disruption to fire-station routines. It also means that installation works can carry on during the winter months when work on heating systems is normally discouraged.

Lessons Learnt – Running Tips

We found that water quality was key. The small water ways on the CHP became blocked very easily if we had poor heating-water quality, even with strainers fitted. We now ensure that all systems are flushed regularly and strainers are in place and maintained.

The close control of the CHPs is vital in getting the most out of the CHPs. To be effective, the CHP **must** be the lead boiler. The controls' system must hold off all other boilers until the CHP is unable to meet the load. They must then be switched off at the earliest opportunity to allow the CHP to continue to run, that is, the boilers should only be used to meet peak loads and then only for short periods. The controls must also give priority to the CHP to heat the hot-water requirements. The 750l buffer vessels help smooth out the peaks and allow the CHPs to run for longer periods without the need for peak-load boilers.

Where there are no sophisticated controls, the CHPs are set to give higher output temperatures than the boilers and DFWH. In this way, the boilers and DFWH see higher return temperatures and assume that the demand is met.

Simple paybacks have been calculated at about 8 years, after the additional gas consumption and maintenance have been taken into account.

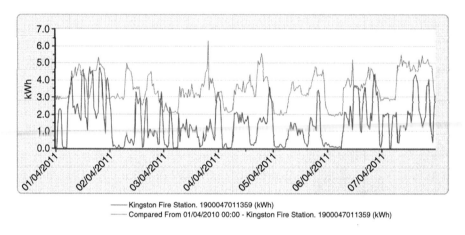

———— Kingston Fire Station. 1900047011359 (kWh)
———— Compared From 01/04/2010 00:00 - Kingston Fire Station. 1900047011359 (kWh)

Figure 1: CHP performance at Kingston Fire Station before installation (blue) and after installation (red).

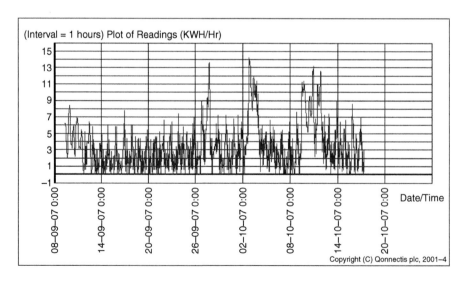

Figure 2: Graph showing that grid-take electricity is zero for several hours per day due to installation of CHP at Battersea Fire Station.

Photovoltaic System at Richmond Fire Station

The installation of a photovoltaic installation on the roof of Richmond Fire Station commenced in November 2003 and was completed in October 2005. The scheme was undertaken in partnership with the Energy Savings Trust who provided a grant of £65,000 towards the cost of the installation. The scheme was managed to a successful conclusion despite major setbacks and this success was attributed to successful partnership working, both across the project team and with the Energy Savings Trust, our major external partner.

The photovoltaic installation was planned to be partly integrated into the roof-replacement and rooflight-renewal works. However, issues were discovered with the internal structure of the roof, and asbestos was identified as a key component in the roof. This discovery limited the amount of photovoltaic cells that could be accommodated on the flat roof. Any cells to be installed on the flat roof would have to be secured with ballast, rather than penetrating the roof to secure a fixing. The original designs were based on securing a fixing into the roof, which was no longer possible.

The outcome of an investigation into the roof structure detailed the load capability of the roof, which permitted a replacement scheme that was to include some photovoltaic cells, located on the sloping roof, in addition to those located and secured on the flat roof using a ballast solution. Acting upon the results of the roof investigation, considerable work was undertaken by the project team and a revised scheme was devised, consisting of fewer cells on the flat roof plus additional cells on the sloping roof at the station.

The 17.5 kWp photovoltaic system was installed both on the flat and sloping roofs at the station, and was also integrated into the replacement rooflights over the appliance bay. This installation produces approximately 11,300 kW hours per year of carbon-free electricity, saving 4.5 tonnes of CO_2 per year, and produces around three-quarters of the station's electricity requirements.

Since the installation at Richmond, a total of 27 sites now have PV installed. The image shows a scheme installed at Park Royal.

Lessons Learnt

Since the project commenced in 2003, there have been a number of areas where hindsight would have delivered a better return.

1. Major specialist contractors (i.e. solar PV installers) probably know more about how to install their systems than do design consultants. Therefore, it would be beneficial to make more use of design and build. This situation is changing slowly as the design consultants gain more experience.
2. It would also make sense to use a larger-framework contract (call-off type contracts) to avoid any tendering for what is essentially the same thing on a repeat basis.

CASE STUDY 2

89 Culford Road: Extreme Low-Carbon Dwelling Refurbishment

Edited by Sunil Shah based on an article written by James Parker and published by BSRIA in December 2009; all photos copyright Roderic Bunn/BSRIA; reproduced by permission of BSRIA

Whilst the focus is on new buildings, it's easy to forget about the refurbishment market. At 89 Culford Road, London, the owners took the opportunity to see what could be done to make existing dwellings low carbon. The house is a three-storey, Victorian town house in the very conservative De Beauvoir Conservation Area of Hackney, north London. The aim was to refurbish an old house in dire need of modernisation and, in so doing, reduce its carbon emissions by 80 per cent. This all sounds quite ground-breaking in the UK, but really it's nothing new. The approach follows the guiding principles of Passivhaus, developed in Germany by the Passivhaus Institut in the 1990s.

The original back wall was completely removed, and rebuilt with reused brick and 200 mm of insulation. Rainwater from the roof is captured by copper drains. Standard, triple-glazed windows were acceptable to the planners as the rear of the house was not governed by conservation rules.

The demolition and rebuilding of Victorian villas and Georgian terraces is not an option. In many areas it would be virtually impossible to get planning permission to do this.

The design involved major work to the fabric of the building, whilst retaining the front façade. Whilst it was never intended to change the structural elevation facing the street, the same cannot be said of the rear of the house, which has been completely rebuilt. There were various motives for this, but notably it was driven by the need to increase and improve the existing accommodation space, which had significantly been reduced by the addition of high levels of thermal insulation and other internal changes. The new rear wall was also built to provide a small extension for the ground and first floors along with a small roof extension.

Whilst the front elevation was kept original, an internal frame was constructed inside the front wall to support the rebuilt floors and also to eliminate thermal bridges. The gap between the existing façade and the internal frame is filled with 140 mm of insulation, with a small, ventilated space between the façade and the insulation. Extra insulation was also added to the floor (100 mm) and to the roof (180 mm).

Airtightness Issues

Surprisingly, for such an old building, a very low air-permeability rate was achieved of 1.1 air changes per hour (ach), prior to plastering, largely due to greater care being taken with wall, floor, window and roof junctions. The first test achieved a result of 5 ach; taping and sealing brought this down to 3.5 ach, with a third and final push (before plastering) bringing it down to 1.1 ach. The final value will be a fraction of that.

However, an airtight house with negligible ventilation would soon become unliveable. Kitchen fumes and humid air from the bathroom would combine with volatile organic compounds released from materials in the house, and with natural smells from occupants, to produce a very unpleasant environment. The high level of airtightness justified the use of a mechanical ventilation and heat recovery unit (mvhr).

This plant, installed on the top floor, will suck out stale air from the bathroom and kitchen, transferring the heat to outside supply air, then distribute it throughout the house via a system of flexible ducts, run down an interstitial gap adjacent to the party wall. As the mvhr manufacturer claims a heat-reclaim efficiency of up to 91 per cent, the fan-power penalty of the mvhr should considerably reduce the dwelling's heating requirements.

Energy Savings

The ventilation strategy was the key feature of the refurbishment in terms of reducing energy demands. Looking at the figures calculated for both before, and after, the work was carried out, it is clear where most of the savings will come from. A reduction of $210\,kWh/m^2$ per annum, or 92 per cent, is outstanding, and would not have been possible without the mvhr–airtightness combination. In fact, this saving is three times the total predicted energy use of the house. The build team is so confident of the thermal capability of 89 Culford Road, that they have installed underfloor heating only on the lower-ground floor, with just two small towel rails in the bathroom to make sure that towels dry out.

The micro, double-glazed windows retained their sash features, albeit replaced with new frames with improved seals (four draught seals per sash) to help with airtightness.

Glazing

In addition to that for the walls, careful choices had to be made for the windows. To complement the rest of the house, they had to have very good seals and very low U-values. This was not easy to fit into a façade in a strict conservation zone. The solution was micro (or slim) double-glazing. These have a small argon-filled gap of only 4 mm. This gave a glazing profile very close to the original. Heat retention is improved with a low-e coating that reflects heat back into the room. The front windows also kept the sash opening, albeit reproduced with improved seals (four draught seals per sash) to help with airtightness. This gives the front of the house a virtually original appearance. The rear of the house, without the strict levels of conservation to be observed, is equipped with a more traditional low-energy window: triple-glazing with double edge seals.

Renewables

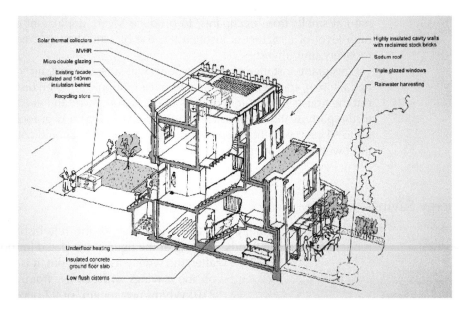

The many extreme low-carbon features of 89 Culford Road; the dwelling has all-new floors, and the staircases were reversed to enable better use of the space.

No self-respecting, low-energy house would look complete without some kind of renewable-energy technology. With this particular development, there was an agreement from the outset to avoid overt 'eco-bling'. The original plan was to have installed a heat-pipe, solar-thermal system. However, to be effective, it would need to have been visible from the street – a consequence of the lack of space to hide it. However, with the introduction of feed-in tariffs, the installation of solar photovoltaics provided an answer. The feed-in tariff reduces the payback period of photovoltaics from 40 years down to just 15 years. The photovoltaics at 89 Culford Road were installed at a cost of around £5,000. They are calculated to provide 1.2 kWp and to meet around 30–40 per cent of the electrical load of the house.

Refurbishment versus New Build

89 Culford Road will be monitored to provide a real measure of its performance. Its lessons can then be rolled out across the domestic sector, with, hopefully, the anticipated energy savings being achieved in practice. The first set of monitoring equipment has been installed by the insulation manufacturer, Knauf.

Sensors have been installed into the walls, particularly on the front wall, and at the interface between different materials to give a temperature gradient through the wall. This should provide an evidence base and allow Knauf to make recommendations as to how its product could be better used.

What 89 Culford Road illustrates is the sheer degree to which the existing domestic-housing stock must be improved if the country is to have any chance of meeting its carbon targets. The effective refurbishment of our existing housing stock is non-negotiable. It's a standard that must apply to all domestic dwellings, not just to the small fraction of new homes that will be built – probably sporadically, given normal economic cycles of growth and recession – over the next 40 years.

Energy End Use	CO_2 kg/yr reduction	% carbon reduction	£/yr reduction	% cost reduction
Space-Heating Gas	4264	91	769	91
Water-Heating Gas	23	3	4	3
Pumps and Fans	−50	−68	−18	−68
Lighting	159	41	57	41
Other Electricity	0	0	0	0
Subtotal	4396	67	812	59
Photovoltaics	516	8	468	
Total	4912	75	1280	93

CASE STUDY 3

Empire State Building to Become a Model of Energy

The retro-fit of the Empire State Building, where a sweeping renovation has been under way since 2007, is expected to cut energy consumption in the 79-year-old, 103-storey building by 38 per cent and save over 105,000 tonnes of CO_2 over the next 15 years.

The Empire State Building is no ordinary office tower. Built in 1931, it draws between 3.5 million and 4 million visitors each year to the Observatory on the 86th floor. At a height of 1472 feet (449 metres), the spire is used for broadcasting by most of the region's major television and radio stations. Its 2.8 million square feet of leasable office space hold a range of large and small tenants, drawn by the building's prestige, its unmatched skyline views and its convenient location at the centre of Manhattan's mass transit system.

Prior to 2008, the building's performance was:

- *Annual utility costs:* US$11 million (US$4/sq. ft.)
- *Annual CO_2 emissions:* 25,000 metric tonnes (22 lbs/sq. ft.)
- *Annual energy use:* 88 kBtu/sq. ft.
- *Peak electric demand:* 9.5 MW (3.8 W/sq. ft. inc. HVAC)

The retro-fit of the Empire State Building (ESB) was motivated by the desire of the building's ownership to:

1) Prove or disprove the economic viability of whole-building, energy-efficiency retro-fits. With a US$500 million capital-improvement programme under way, ownership decided to re-evaluate certain projects with cost-effective, energy-efficiency and sustainability opportunities in mind. Energy-efficiency and sustainability measures provide amenities (lower energy costs, easier

carbon reporting, daylighting etc.) that set the building apart from surrounding tenant space.

2) Create a replicable model for whole-building retro-fits. There are known opportunities to cost-effectively reduce greenhouse-gas emissions, yet few owners are pursuing them. ESB ownership wants to demonstrate how to cost-effectively retro-fit a large, multi-tenant, office building to inspire others to embark on whole-building retro-fits.

Identify opportunities	Evaluate measures	Create packages	Model iteratively
• 60+energy-efficiency ideas were narrowed to 8 implementable projects	• Net present value	• Maximize net present value	• Interative energy and financial modelling process to identify final eight recommendations
• Team estimated theoretical minimum energy use	• Greenhouse-gas savings	• Balance net present value and CO_2 savings	
• Developed eQUEST energy model	• Dollar to metrictonne of carbon reduced	• Maximize CO_2 savings for a zero net present value	
	• Calculated for each measure	• Maximize CO_2 savings	

3) Reduce greenhouse-gas emissions.

'The goal with ESB has been to define intelligent choices which will either save money, spend the same money more efficiently, or spend additional sums for which there is reasonable payback through savings. Addressing these investments correctly will create a competitive advantage for ownership through lower costs and better work environment for tenants. Succeeding in these efforts will make a replicable model for others to follow.' – Anthony E. Malkin

Project Development Process

Using ESB as a convening point, a collaborative team was formed to develop the optimal solution through a rigorous and iterative process that involved experience, energy and financial modelling, ratings' systems, technical advice, and robust debate. Key points include:

1) Five key groups and a host of contributors used a collaborative and iterative approach.
2) A 4-phase project-development process helped guide progress.
3) A variety of complementary tools were used and developed to triangulate to the best answer.

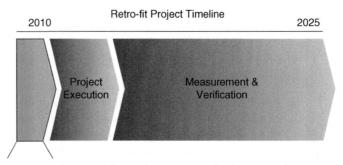

Project development is focused on understanding current performance, analysing opportunities, and determining which projects to implement.

Projects

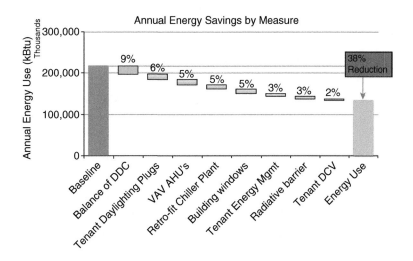

Annual Energy Savings by Measure

Examples of projects that have been undertaken include:

Retro-fitting the chiller plant was necessary to account for the reduced cooling demand. The electric chiller plant was completely rebuilt with only the heat exchanger shells reused. Efficiency gains from 0.8 kW/tonne to 0.55 kW/tonne at full load and proportionately better at part loads, increasing available cooling capacity and reducing capital expenditure and operating expense.

Anticipated savings: Carbon: 1,430 tonnes/yr; Energy: 11.4 MkBtu/yr; Cost: US$676 k/yr

All 6,514 windows have been retro-fitted where a window refurbishment processing centre was built on site at ESB, reducing transportation emissions and creating 50 local union positions. Each window was removed and remanufactured, involving the insertion of a thin film and gaseous mixture between the existing two panes of glass. 95% of the glass was reused, with the insulating value being increased from R2 to R7. The refurbishment reduced overall cooling and heating expenses.

Anticipated savings: Carbon: 1,150 tonnes/yr; Energy: 11.4 MkBtu/yr; Cost: US$410 k/yr

The Building Management System (BMS) was upgraded and sub-meters were installed on each floor including replacing outside air dampers that are now controlled by CO_2 sensors reducing both cooling and heating requirements whilst providing the required fresh air.

The Tenant Energy Management Program helps tenants to better understand how energy is being used. It provides sub- meters, installed for individual tenants to allow monitoring of electrical energy consumption, available for building management as well as tenant energy management. Information will be available on-line to allow tenants to benchmark themselves against other tenants.

Anticipated savings: Carbon: 743 tonnes/yr; Energy: 6.9 MkBtu/yr; Cost: US$387 k/yr

Installation of reflective insulation behind each of the 6,514 radiators involved radiative barriers being insulated, redirecting heat inward, reducing the project's use of steam and, thus, operating costs. Currently one steam radiator is located beneath each window in the Empire State Building, with almost half the heat lost through the wall.

Anticipated savings: Carbon: 480 tonnes/yr; Energy: 6.9 MkBtu/yr; Cost: US$190 k/yr

Maximising daylighting through designing more energy-efficient lighting systems and lighting controls for tenant spaces. Tenant plug loads for computers and monitors were managed through plug load-occupancy sensors.

Anticipated savings: Carbon: 2,060 tonnes/yr; Energy: 13.7 MkBtu/yr; Cost: US$941 k/yr

Replacement of 300 existing air-handling units with fewer and more efficient units to provide variable air volume delivering greater occupant comfort.

Anticipated savings: Carbon: 1,520 tonnes/yr; Energy: 11.4 MkBtu/yr; Cost: US$703 k/yr

Upgrading existing and installing new building controls helps to optimise HVAC system operation as well as providing more detailed sub-metering of electricity use.

Anticipated savings: Carbon: 1,900 tonnes/yr; Energy: 20.6 MkBtu/yr; Cost: US$741 k/yr

Demand control ventilation to provide the appropriate amount of outside air into the building through the use of CO_2 sensors in occupied spaces. This approach saves the energy needed to pre-heat or cool the outside air, together with ensuring that air-quality standards are met.

Anticipated savings: Carbon: 300 tonnes/yr; Energy: 4.6 MkBtu/yr; Cost: US$117 k/yr

Project Description	Projected Capital Cost	2008 Capital Budget	Incremental Cost	Estimated Annual Energy Savings*
Windows	$4.5m	$455k	$4m	$410k
Radiative Barrier	$2.7m	$0	$2.7m	$190k
DDC Controls	$7.6m	$2m	$5.6m	$741k
Demand Control Vent	Inc. above	$0	Inc. above	$117k
Chiller Plant Retro-fit	$5.1m	$22.4m	–$17.3m	$675k
VAV AHUs	$47.2m	$44.8m	$2.4m	$702k
Tenant Day/Lighting/Plugs	$24.5m	$16.1m	$8.4m	$941k
Tenant Energy Mgmt.	$365k	$0	$365k	$396k
Power Generation (optional)	$15m	$7.8m	$7m	$320k
TOTAL (ex. Power Gen)	$106.9m	$93.7m	$13.2m	$4.4m

This package of measures also results in enhanced indoor environmental quality and additional amenities for tenants:

- *Better thermal comfort resulting from better windows, radiative barrier, and better controls;*

- *Improved indoor air quality resulting from DCV;*
- *Better lighting conditions that coordinate ambient and task lighting.*

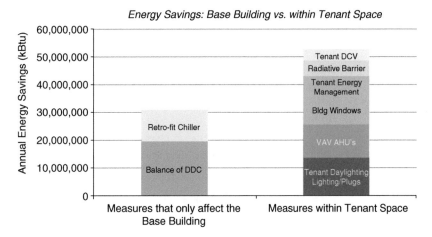

Energy Savings: Base Building vs. within Tenant Space

Lessons Learnt

In the process of developing specific project recommendations, a number of key lessons for the retro-fit of large, multi-tenant, commercial office buildings have been identified.

1. **Developing robust solutions requires the coordination of several key stakeholders**

   Planning energy-efficiency retro-fits in large, commercial office buildings must address a dynamic environment, which includes changing tenant profiles, varying vacancy rates, and planned building renovations. In the Empire State Building, the project team included engineers, property managers, energy modellers, energy-efficiency experts, architects, and building management.

   Each of these stakeholders was needed to help build a robust energy model that addressed the building's changing tenant profile and helped the team to model the impacts of its energy-efficiency strategies. Coordination also included the tenants. Involving tenants and considering their perspective at an early stage is critical because more than half of the energy-efficiency measures that will be implemented at the Empire State Building involve working both with tenants and within their spaces.

2. **Maximising energy savings profitably requires planning and coordination**

   For an energy-efficiency retro-fit to be cost-effective, the retro-fit needs to align with the planned replacement or upgrades of multiple building systems and components. For instance, the Empire State Building had plans under way to replace its chillers, fix and reseal some of its windows, change corridor lighting, and install new tenant lighting with each new tenant.

   Since these upgrades were already going to be carried out, the team redesigned, eliminated and created projects that cost more than the initial budget but achieved significantly higher energy savings over a 15-year period. When these energy savings were accounted for, along with the added upfront project costs, the net present value of the energy-efficiency retro-fit projects was better than that of the initial retro-fit projects. However, the energy savings are not substantial enough to offset the full capital cost. This means that carrying out energy-efficiency projects long before major systems and components are ready for replacement will likely be cost-prohibitive, with a poor net present value. The large volume of existing commercial buildings suggests that there is a tremendous opportunity to reduce carbon emissions from existing buildings through energy efficiency; however, capturing these reductions in a profitable manner demands careful planning and coordination to ensure that energy-efficiency retro-fits align with building replacement cycles.

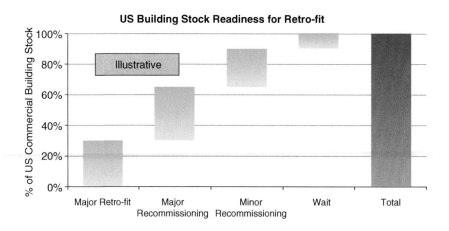

For many buildings that are not undergoing, or approaching, major replacements, there might still be a major opportunity to retro-commission the building. Retro-commissioning improves the operation of existing buildings, many of which are typically run to minimise complaints, rather than optimise energy performance and create comfortable working environments. Retro-commissioning can typically reduce energy use between 5 and 15 per cent in most existing buildings. Developing a tool, or set of tools, that can quickly assess a building to determine whether it is a candidate for a whole-building retro-fit, retro-commissioning or requiring no action until a few years later will dramatically improve the effectiveness of funds and efforts directed towards energy-efficiency retro-fits of existing buildings.

3. *Tension between business value and reducing CO_2 emissions*
   In the Empire State Building, maximising profitability from the energy-efficiency retro-fit leaves almost 50 per cent of the CO_2 reduction opportunity on the table. The building owner, whilst still selecting an optimal package of measures with a high net present value, sacrificed 30 per cent of profit to deliver more CO_2 reductions and improve the lighting and tenant comfort within the building. Changes in energy prices and/or the cost of energy-efficiency technologies may help to better align profit maximisation and CO_2 reduction. However, as things stand currently, there is a gap between the socially desirable amount of CO_2 reduction and the financially beneficial amount of CO_2 reduction from a building owner's perspective.

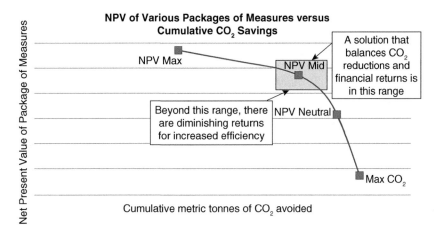

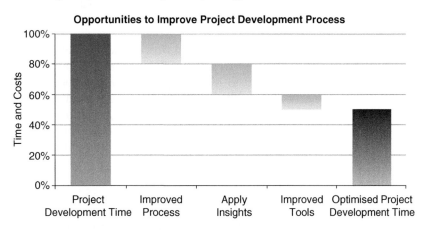

4. *Creating a replicable process for a whole-building retro-fit*

   Developing the energy-efficiency strategies that will be implemented in the Empire State Building took over nine months of intensive building audits, brainstorming sessions, energy modelling, documentation, and financial analysis. Although the Empire State Building is a unique building with unusual challenges, the process used to drive deep energy and carbon savings in the Empire State Building can be made much more efficient. Having completed the recommendations for the Empire State Building, the project team recognises a number of opportunities for condensing the study period: developing experienced teams; creating tools for rapidly diagnosing and categorising a building (or a portfolio of buildings); quickly developing a 'first-cut answer'; and developing and using tools to quickly iterate between financial and energy modelling to arrive at the optimal package of measures.

5. *Carbon regulation does not significantly affect the Empire State Building recommendation*

   The financial decision-making tool helped the team to understand that the recommended package of energy-efficiency measures would not significantly change even if there were to be carbon regulation that leads to higher energy prices over time. Carbon regulation that changed energy prices by less than two per cent per year had little effect on the financial performance of the modelled packages. However, if energy prices rise by over 8 per cent (associated with a carbon price of approximately US$30/metric tonne of CO_2), a package with all of the energy-efficiency measures that were analysed (as opposed to those that were recommended) rises to NPV neutral instead of NPV negative.

Project Finances

So how much does all this cost? Below are answers to some common questions regarding cost, savings, the performance contract, and financing.

1. ***How much does the recommended package of eight energy-efficiency projects cost?***

The total incremental cost for efficiency beyond planned infrastructure upgrades is US$13.2 million. Some funding within the existing capital budget was reallocated to different projects, whilst some funding was removed as projects were deleted, and other funding was added to support new efficiency projects. The sum of all these changes, deletions, and additions is US$13.2 million. This incremental cost includes the soft and hard costs required to provide, install, and set into proper operation all equipment and systems that make up the project.

2. ***How much money does this package save?***
The package of eight measures saves US$4.4 million annually once fully implemented.

3. ***What is the total value of the performance contracts?***
Five of the eight projects will be implemented using a performance contract. The total cost (not incremental) to implement those measures is US$20 million. These projects will save US$2.4 million of the total US$4.4 million annual savings.

4. ***How will the project be financed?***
A financing solution is currently being sought that covers the funding of all costs, and spreads all payments over the performance period or term.

Implementation-period draws will be required monthly through the 18-month implementation schedule to meet construction requirements. The owner is seeking project financing that can be paid back over a 15 to 20-year period, depending on the offers available from respective financial institutions.

Project Benefits

The project not only saves money for the Empire State Building and for tenants, but it has other benefits as well.

Occupant Benefits
Projects in the recommended package, including the windows, radiative barrier, and better controls, will provide enhanced thermal comfort to building occupants. Demand control ventilation will improve indoor air quality by ensuring that adequate, though not excessive, ventilation is provided to tenant spaces. A high-quality visual environment will be provided through tenant lighting and daylighting efforts that layer ambient, direct/indirect, and task lighting to maximise comfort whilst saving energy.

Chiller Plant Retro-fit and Sizing
The recommended package of retro-fit measures provides a ~30% reduction in the amount of heat that needs to be removed from the building. This cooling load reduction, along with the existence of industrial chillers within the Empire State Building, enabled the project team to recommend retro-fitting the chiller plant, rather than installing a new chiller plant, to handle additional capacity. By retro-fitting the chiller, the construction of a new chiller plant for the Empire State Building can be delayed. By delaying the need for a new chiller plant, the Empire State Building will also be able to downsize its new plant based on the load reductions the building can realise through efficiency measures over the next 15 years and beyond.

Electrical Demand

The recommended package of measures also reduces the building's peak electrical demand by approximately a third, freeing up currently constrained electrical capacity.

CO_2 Reduction

Based on the phasing-in of the recommended projects, it is estimated that over 100,000 metric tonnes of carbon will be saved over the next 15 years. This is equivalent to CO_2 emissions from the use of approximately 11 million gallons of gasoline.

The ability to replicate their model has been amongst the guiding principles of the project. To that end, the measurement, performance-modelling and financial tools, and other material developed in the analysis process by the project partners, are all being made available on-line for public use at www.esbsustainability.com and www.esbnyc.com

Endnotes

1 Adapted from 'Embodied Carbon: The Inventory of Carbon and Energy (ICE); Professor Geoffrey Hammond and Craig Jones; BSRIA, 2010

2 Redefining Zero: Carbon Profiling as a Solution to Whole Life Carbon Emission Measurement in Buildings, RICS Research, May 2010

3 Birkland, J., 2002, Design for Sustainability: A Sourcebook of Integrated, Eco-Logical Solutions, Sheffield, Earthscan Publications

4 BCIS, 2006, Life Expectancy of Building Components, 2nd ed., London, Connelly-Manton (Printing) Ltd

5 Cole, R. J., Kernan, P. C., 1996, Life-Cycle Energy Use in Office Buildings. Building and Environment, vol. 31, no. 4. Oxford, Elsevier

6 M&S Carbon Trust Zero Carbon Spec Case Study, Deloitte dcarbon8, September 2010

Part 4

Environmental Areas

The delivery of sustainable refurbishments is much greater than carbon issues alone, and covers both environmental and social factors as well. The ability to introduce environmental benefits into refurbishments can help to improve the valuation of buildings by increasing their sustainability-performance rating of BREEAM or LEED.

The table below captures the sustainability benefits that can be delivered from the various environmental measures discussed in this part. These benefits cross over a number of areas and are rated according to their number of stars – the greater the number of stars, the greater the benefit.

Sustainability Category	Material Use & Resources	Water Conservation	Biodiversity	Transport
Management	**	**	**	**
Emissions to Air	*		*	***
Land Contamination	*		*	
Workforce Occupants	*	*	**	*
Local Environment & Community	**	*	**	**
Life cycle of building/ products	***	***	*	
Energy Management	**	**	**	*
Emissions to Water	*	***	**	
Use of Resources	***	*	**	*
Waste Management	***		*	
Marketplace	*	*	*	*
Human rights	*	*	*	*
Biodiversity		*	***	
Transport	*			***

Sustainable Refurbishment, First Edition. Sunil Shah.
© 2012 Sunil Shah. Published 2012 by Blackwell Publishing Ltd.

12 Material Use and Resource Efficiency

What is a sustainable material? Timber is a sustainable material. If you cut down a tree to use in making something, then you sustain the supply by planting another tree. A sustainable material, or a sustainable resource, is something whose production is supported indefinitely by nature. It is not always the same thing as a renewable resource.

Sustainable building is an essential aspect of widening efforts to create an ecologically responsible world. A building that is sustainable must, by nature, be constructed using locally sustainable materials: that is, materials that can be used without any adverse effect on the environment, and which are produced locally, reducing the need for their transportation.

The ability to achieve this can be challenging where cost constraints, material availability and conflicting certification schemes can make the inclusion of sustainable materials an uphill struggle.

Chapter Learning Guide

This chapter will provide an overview of the identification and implementation of sustainable materials.

- Introduction to sustainable materials covering the drivers for, and barriers to, their uptake and integration;

(Continued)

Sustainable Refurbishment, First Edition. Sunil Shah.
© 2012 Sunil Shah. Published 2012 by Blackwell Publishing Ltd.

- Review of material-certification schemes for a range of product types or specific materials;
- Approach to incorporating sustainable materials into the design process for major refurbishments;
- Checklist of minimum requirements and good practice to implement sustainable-material refurbishments.

Key messages include:

- Material-certification schemes should be treated with caution – look at the level of life-cycle assessment performed;
- Recycled content and recovered materials can easily achieve 10 per cent of the total construction materials.

12.1   Introduction

Materials used for sustainable refurbishments can be grouped under four main types:

1. Traditional materials used in a traditional way – for example, solid timber for floors and windows;
2. Traditional materials used in an innovative way – for example, glulam beams;
3. Natural crops and fibres used as replacements or alternatives – for example, straw bales for insulation and walls;
4. Innovative materials – for example, modern fibre composites with natural fibres.

There are, however, a number of barriers to greater take-up and use of sustainable materials, despite the fact that there are some drivers beginning to appear that will encourage the use of these materials (see Figure 12.1). These barriers include:

- **Market Barriers** based on the availability of the materials at the time when they are required, together with perceptions and psychological barriers on the part of the designer and end-user. The end-user barriers are particularly significant as they look at the typical change issues associated with many other sustainability criteria. These will perhaps be the hardest barriers to overcome as they are based on the expectations and attitudes that customers, tenants and occupiers have with regard to their space.
- **Product Information and Quality** including statutory performance covering building regulations and standards together with the durability of the product. Where information is available, the clarity of the data provided is also necessary to allow an understanding of the product, its installation and its benefits. Unfortunately, in this latter respect, the information available from manufacturers on their websites is not consistent in its provision of performance information to allow a comparison to be made.

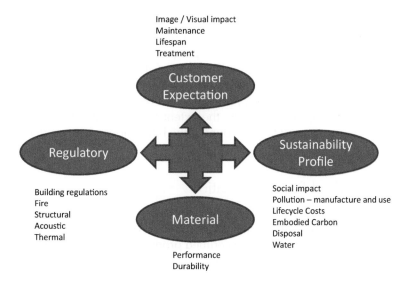

Image / Visual impact
Maintenance
Lifespan
Treatment

Customer
Expectation

Regulatory

Sustainability
Profile

Building regulations
Fire
Structural
Acoustic
Thermal

Social impact
Pollution – manufacture and use
Lifecycle Costs
Embodied Carbon
Disposal
Water

Material

Performance
Durability

Figure 12.1: Barriers and drivers to the take-up of sustainable materials.

- **Regulatory concerns** regarding whether the materials comply in the longer term, without early degradation, with building regulations, with standards for acoustic or thermal performance, or with standards for lighting levels. There needs to be a belief that the expected performance of the material will be achieved – this is particularly difficult for new products on the market.
- **Sustainability Assessment** of the product capturing any commendations or certifications that the material has received. There is a need to understand the environmental impact of a material, which embraces its life-cycle issues related to its manufacture, use and disposal; this amounts to an assessment of more than just the embodied-carbon issue. Included within this assessment are the whole life-cycle costs including those for repairs and maintenance. To move further and identify the pollution and social costs is a relatively new activity and one that will not enable a true comparison of materials.

12.2   Material-Certification Schemes

An increasing number of manufacturers are making environmental claims about products, which can make it difficult for consumers to compare materials. One procedure that can help consumers to combat 'greenwash' is certification. There are a number of highly reputable programmes that enable the certification of particular products, but they do vary in the level of assurance that they can provide. Therefore, it is important to know which certification programmes are reliable and why.

The most extensive, reliable, and expensive certification programmes perform a thorough life-cycle assessment of products or services. The most common form of 'greenwashing' is the 'sin of the hidden trade-off'. In such a case, a company makes a claim about the environmental benefits of one specific aspect of a product (for example, it is made from recyclable materials), whilst failing to disclose that other aspects of the product might be highly detrimental to the environment (for example, it might be manufactured in a very energy-inefficient manner, relative to that of competing products).

Life-cycle assessment programmes are able to evaluate the environmental impact of the entire life cycle of the product. The most ecologically beneficial products are designed at the outset so that their impact on the environment at each of their life-cycle stages will be minimal. Life-cycle assessment programmes attempt to determine the extent to which this is this case. All life-cycle assessments must take account of the boundaries of the production system that is being evaluated[1].

Independent Third-Party Verification

As part of the process of performing life-cycle assessments of products, reliable certification evaluations are carried out by a third party, rather than by the companies who sell the certified products, so that the evaluations are protected from conflicts of interest. Reliable programmes include input from a broad range of stakeholders, such as consumer groups, environmental groups, businesses, and government representatives. They also adhere to standards established by widely recognised, independent organisations, such as the American National Standards Institute (ANSI) and the International Organization for Standards (ISO). The most reliable certification programmes conduct on-site inspections of facilities and perform their own independent testing of the products that they certify. They also require annual follow-up assessments in order for certification to be maintained.

Green Seal, Ecologo, MBDC cradle to cradle, BRE Green Guide, and Green Building Council of Australia are amongst many who offer life-cycle-assessment certification for a broad range of products, details of which are then listed on their websites. These programmes comprise life-cycle assessment, on-site inspection, and independent testing to ensure that companies and products meet their standards. Co-op America, a non-profit organisation, provides a screening process that is less robust than the above-mentioned certification programmes, but is a much more affordable option for smaller businesses. In order to be listed in the Co-op Green Pages Directory, businesses must complete a questionnaire and provide answers that demonstrate to a screening committee a commitment to social and environmental responsibility.

There are a number of programmes that provide life-cycle-assessment certification for buildings. LEED and BREEAM have the largest international programmes. Green Guard provides certification that products are not detrimental to healthy, indoor air quality.

The Fairtrade Labelling Organizations International is an international non-profit association of fair trade organisations that have created common standards for certifying fair-trade products. Fair-trade certification ensures that products were created by companies that deal directly with the people who produced them so as to ensure a fair price for the product. Fair-trade organisations must meet environmental standards as well as provide wages and services that enable producers and their families to meet basic needs for health and education.

The Forest Stewardship Council (FSC) is one of many organisations that certify that wood and paper products were created in a manner that does minimal damage to forest ecology and to neighbouring local economies.

Although legitimate and reliable certification is an important procedure for guarding against 'greenwashing', it does have its limitations and drawbacks. One of the greatest limitations is that there are a number of green products and processes that are not yet covered by highly reliable, life-cycle-assessment certification. It remains to be seen whether innovation in certification programmes can keep pace with the rapid, daily development of new, green products and services. The other main limitation imposed by certification is cost. As a general rule, the more thorough, informative and reliable a certification process is, the more expensive it will be. Costs of £15,000 (in the UK) for an assessment per product to gain certification, followed by annual costs for monitoring, add up to a significant cost that many small or start-up businesses cannot afford.

Product Certification Project – Green Building Council of Australia (GBCA)

The Assessment Framework for Product Certification Schemes was released in June 2009, as part of the GBCA's ongoing review of the Green Star environmental-rating system for buildings.

The Framework is the result of extensive dialogue with Green Star stakeholders, the appointment of an independent Expert Reference Panel and feedback from manufacturers and building-industry professionals.

The GBCA has developed the new Assessment Framework to:

- clarify best-practice benchmarks and establish expectations for manufacturers and suppliers of fit-out products, as well as for the certification schemes that are recognised through Green Star;
- reduce the costs of product certification for manufacturers and suppliers;
- reduce the costs and remove the commercial barriers associated with Green Star fit-outs.

Products and materials that are certified by GBCA-recognised certification schemes will have greater access to the 'deemed to satisfy' compliance status within the Green Star Material Calculators.

http://www.gbca.org.au/green-star/materials-category/product-certification-project/2292.htm

12.3 Material Procurement

Many mainstream products commonly used in construction, such as blocks, bricks and boards, contain significant amounts of material that have been recovered from the waste stream at no extra cost or risk whilst meeting quality standards. For example, different brands of plasterboard may contain between 15-per-cent and 99-per-cent recycled content. It is not always obvious that new building products contain such material – but, by selecting appropriate products, contractors can be more efficient in their use of material resources without compromising on cost, quality or tight deadlines. Specifying recycled materials is a strategy to pursue alongside established options for reclaiming and reusing products (such as partitions and ceiling grids). It is a simple way of delivering quantifiable environmental benefits – without compromising on the aesthetic qualities or technical-performance requirements that are demanded by a particular scheme. Targets of 20-per-cent recycled content should be set[2].

Sustainable Material Procurement Approach:

1. Collate a list of the main materials to be used in construction – for example, concrete, glass, aggregate, cladding;
2. Set targets for sustainable, locally sourced and recycled content materials;
3. Confirm that standard sizes are to be specified, for example, for windows and partitioning;
4. Calculate locally sourced and recycled content alternatives to traditional sources for the top-ten materials (based on cost);
5. Specify requirements within tender process, including the need to verify sourcing of materials, for example, FSC timber;
6. Review responses against targets set and confirm chain of custody for sustainable materials such as timber.

Building Research Environments, (BRE) Green Guide, The American Institute of Architects' Environmental Resource Guide and Green Building Specification provide a remarkably thorough compilation of the environmental aspects of scores of materials. LEED and BREEAM provide some synthesis, incorporating materials-specific standards such as Forest Stewardship Council certification for sustainably harvested wood, low VOC emissions for interiors, and defining criteria for recycled content or regional sourcing.

A typical list of materials to avoid includes:

- Timber from non-Forest Stewardship Certification (FSC) scheme;
- Ozone-depleting substance as a coolant;
- Insulation materials with a global warming potential above 5;
- High volatile-organic-compound (VOC) content paints, varnishes and finishes;
- Lead-based paints;
- Flooring that will release VOCs;
- Asbestos;
- Peat and weathered limestone;
- High embodied-energy materials.

12.4   Designing-in Sustainable Materials

Specification is an obvious area where savings can be made – in environmental impact as well as in terms of the usual requirement to reduce cost. The immediately obvious example lies in choosing the most sustainable or renewable materials wherever possible; however, there are other important considerations, including choosing the best types of materials for the job in terms of their thermal efficiency and their properties ('Is concrete being specified here because it is the

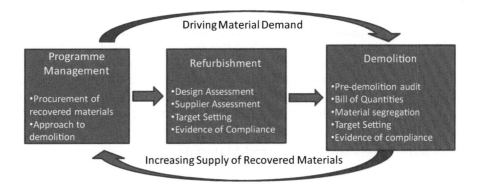

Figure 12.2: The ICE demolition protocol (adapted from ICE).

best material for the particular circumstances, or because it is easily available and commonly used?').

For any major refurbishment where demolition takes place, the potential reuse of materials should be fully assessed at design stage to cover the following areas as part of any demolition protocol (see Figure 12.2)[3]:

- Assessment of design for potential to specify recovered (recycled/reclaimed) materials;
- Assessment of supply chain to provide cost-effective recovered materials;
- Target set for procurement;
- Evidence provided of compliance with targets.

Timber should be chosen specifically for its purpose; different woods have different properties and some will be suitable for structural work whilst others will be better used for windows and doors, or for cladding. Many people in the UK specify imported cedar for cladding, but there are native woods that are equally suitable for this purpose, provided that the architectural detailing is appropriate: larch, for example, is excellent as a cladding material and grows well in temperate climates. By specifying native woods, designers help to create jobs in forestry and reduce the environmental impacts of transporting timber between countries.

Where timber is imported, it is important to ensure that it is from a sustainably managed source. FSC certification is one way to accomplish this. Insulation is another area in which sustainable materials are readily available. There are many products ranging from sheep's wool to cellulose-based insulation, many now well tried and tested. Paints, varnishes and other finishes should, wherever possible, be water-based, rather than solvent-based. This is because solvent-based products are generally more energy-intensive in their manufacture, are derived from oil, which is a finite resource, and give off volatile-organic-compound (VOC) emissions in use.

Life-cycle costing is an important principle to bear in mind when choosing materials: repair and maintenance implications have a major effect on the lifetime impact (and financial cost) of a building or structure.

However, one of the most important principles in ensuring sustainability in the materials chosen is that of designing a built asset so that its materials, structure and services work in harmony to deliver the design criteria. For example, timber is widely considered to be a sustainable material: but if it is used as a cladding for a concrete-frame building that is only intended to have a short design life, then resources are being used inappropriately and, therefore, are wasted. By contrast, a high-thermal-mass building with a concrete frame that uses the properties of the material in its heating, cooling and ventilation strategies, with a design life of sixty years, is likely to reach very high sustainability standards.

Recycled Content and Recovered Materials

Construction and demolition activities produce significant volumes of waste each year – for example, some 60 million tonnes of waste each year in the UK[4]. Whilst much is recycled, mainly into construction applications, the potential exists to recover and reuse much more material, both from construction and demolition activities and from other waste streams.

The Waste and Resources Action Programme (WRAP) provided a study that found that 10 per cent of the materials' value of any construction project could easily derive from recycled content *at no extra cost*[5]. In order to achieve its maximum effectiveness, this requirement should be included in the contract specification; ideally, it should also be a requirement placed on designers from the very beginning of the scheme, so that maximum opportunities are available. The 10-per-cent-by-value stipulation would apply to the project as a whole, not just to individual products.

The contractor should have complete flexibility to find the most cost-effective solution, depending on the project type and requirements, and on local availability: one option for an office project might, for example, involve a large volume of low-value, recycled aggregate, together with a small volume of some high-value product such as a decorative finish; alternatively, a good, local supply of reclaimed brick for a particular part of a building might be a better option for a similar scheme elsewhere in the country.

If demand for recycled content in construction products can be increased by setting value-based requirements that are achievable in everyday practice, this will increase the value of recovered materials. In turn, this will mean that segregating and recycling materials will become much more financially attractive than disposal to landfill – currently the 'easy option' – particularly as disposal costs continue to rise.

12.5   Material Resource Efficiency

The refurbishment of a building can have an immediate impact on the quantities of solid waste produced at the construction stage. Making the most efficient use of materials in the design, and designing to standard sizes with minimum cutting of materials, can be achieved within almost all project types and budgets, and these measures may have the most impact on the waste produced during construction. In some cases, it might be appropriate to explore the possibilities of using prefabricated or pre-assembled materials or modules; in other schemes, particularly large-scale projects, inviting key suppliers to take part in design and sustainability workshops can bring substantial benefits (see Figure 12.3). Resource efficiency involves:

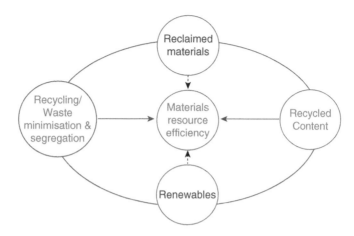

Figure 12.3: Key components of material efficiency.

- Reusing materials for the same purpose where possible (ornamental stonework, bricks);
- Reusing materials on site where possible or from other sites (recycled aggregates);
- Using manufactured products with a high recycled content (tiles, plasterboard, blocks, chipboard);
- Using secondary materials in preference to primary materials (PFA and GGBS as replacements for cement);
- Minimising transport distances, emissions and energy use over whole life of material.

Modular designs tend to waste less material, and need not restrict design flair to a prohibitive extent if used appropriately and in context. However, even with modular designs, it is important to consider component sizes, if materials are to be used to best effect. For example, if designing a building that has blockwork internal walls, consider:

- Internal walls designed to block lengths;
- Block-sized openings;
- Coursing to avoid cut blocks;
- Lintels the same size as blocks, with no cuts.

Choosing the right material for the right purpose is essential in terms of material efficiency – ensuring that materials are used appropriately for their properties, not simply because they have commonly been used in the past and are familiar, and therefore easy to specify.

Methods of assembly and their associated fastenings can have a major impact on the feasibility of building 'deconstruction' and the reuse of components, whether on-site or off-site: bolting steel frames, rather than welding them, for example.

Certain materials are difficult to recycle because of their very nature. Composites are often difficult to separate into their component parts – a composite cladding panel might consist of decorative and protective coated metal, adhesive and foam insulation – and, therefore, recycling would not be feasible. Each of the components might be recyclable if used independently: this is a more efficient use of the materials.

12.6   Site Waste Management

Site-waste-management plans (SWMPs)[6] provide a framework to improve environmental performance, meet regulatory controls and reduce the rising costs of disposing of waste. Adopting a site-management approach, based around an effective SWMP, can bring many benefits, including:

- Better control of risks relating to the materials and waste on site;
- Provision of a demonstration of waste management and cost and risk control;
- Compliance with legislation;
- A framework to make cost savings through better management of materials' supply, materials' storage and handling, and better management of waste for recovery or disposal.

Waste Minimisation in Office-Refurbishment Projects: an Australian Perspective[7]

Building stock in many cities can be described as 'mature', that is, either refurbished some time ago or reaching a stage where major refurbishment is necessary. All refurbishment, however, generates some amount of solid waste and generally this is at a higher rate than that of new construction for a given floor area, much of which is potentially reusable or recyclable.

There are several commonly cited impediments to waste minimisation in general construction projects, including: available space and time restrictions that have been shown to limit on-site sorting of the waste stream; work practices and attitudes that might mitigate against reuse and recycling; small quantities of a recyclable material that might be uneconomical to sort and transport to a recycling facility. It is likely that several of these problems may be heightened in the more restricted area of refurbishment projects where specific management skills are needed.

A series of expert practitioners involved in the design, specification and construction of refurbished buildings provided responses in the four categories of building fabric, fittings, finishes, and services' components (see Figures 12.4 to 12.7).

The building fabric removed in a commercial refurbishment project is likely to receive a significant level of recycling at present. Almost all of this recycling happens off site. Aluminium, structural steel, and steel reinforcement are reportedly recycled at the rate of 86 per cent, 79 per cent and 84 per cent respectively. Heavy masonry materials, such as bricks, blocks and concrete, are also commonly recycled (rates of over 70 per cent for each element). The only element of the building fabric whose prime destination is landfill is that of the stairs, this being probably due to their highly customised nature.

Landfill was the principle destination reported for most fittings removed from refurbishments, except for suspended ceilings, partition walls, workstations and glazed partitions. Workstations were commonly reused both on site and off site (35 per cent for each category). Very little recycling was reported for fittings.

The majority of all finishes removed during refurbishments end up in landfill and no recycling on site was reported. Reuse for carpets is reportedly a growing area. Plasterboard recycling was an area of considerable disagreement amongst the experts. Whilst several reported that no recycling occurred, a few were able to report high levels of recycling. The differences appear to be location-based with recycling facilities being widely available in Victoria whilst very little plasterboard recycling occurred in other states.

Finally, high levels of recycling off site occur with most services' components but there was very little reuse reported. Recycling facilities are available for most services' components but refrigeration components appear to lag behind in this respect.

Case Study Project – Preliminary Findings

The waste outcome from the refurbishment of a 22-storey, government-office building in Sydney was tracked. The building was constructed in 1979 and had undergone no major refurbishment since that time. The building was to remain continuously occupied during the refurbishment and, consequently, the project was staged over a five-year period.

Findings from the project have identified three key areas. Firstly, the presence of asbestos insulation in the inter-floor and duct spaces in the building severely constrained the scheduling of the refurbishment and limited the amount of materials' recycling that ended up being done. Secondly, the continued occupation of the building during refurbishment had the result of stretching the project's progress over a long period of time. Major work had to be done in short bursts over the holiday periods and there was very little opportunity for on-site sorting or for storage of items for later reuse elsewhere in the building. Thirdly, owing to scheduling difficulties arising from the need to accommodate the continued operation of the building and the continued public access, the decision was taken to break a very large project into discrete small contracts for the various stages of the work. This meant that there was little incentive for individual contractors to sort, store and salvage materials in small quantities.

A structured approach to the implementation of SWMP from pre-construction through to on-site use and validation via key performance indicators is described in Figure 12.8.

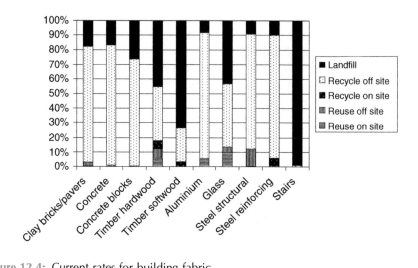

Figure 12.4: Current rates for building fabric.

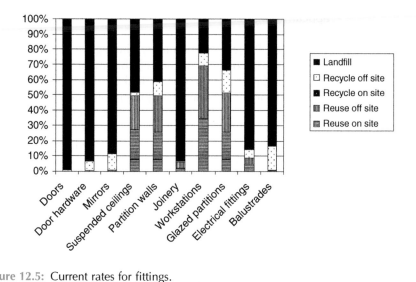

Figure 12.5: Current rates for fittings.

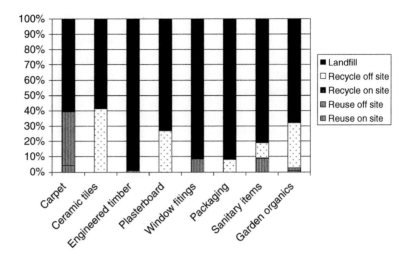

Figure 12.6: Current rates for finishes.

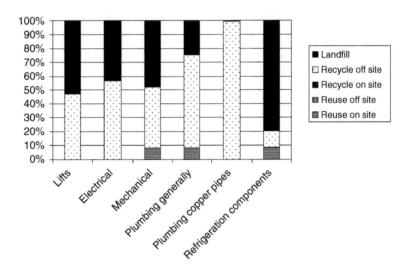

Figure 12.7: Current rates for services.

Project Stage	SWMP Section
Policy and Set-up	Identified nominated owner
Preparation and concept design	Record waste prevention actions
Detailed design	Forecast waste
	Record waste prevention actions
Pre-construction	Specify waste carriers
	Plan waste destinations
	Record waste management and recovery actions
Construction	Record actual waste movements
Post completion and use	Assess performance against KPIs

Figure 12.8: SWMP approach.

12.7 Materials and Resource Efficiency Checklist

Table 12.1 provides a summary of the items to be discussed during a sustainable-refurbishment project, capturing where specific items are necessary, recommended or nice-to-have options.

Table 12.1: Materials and resource efficiency checklist

Minimum Standard	Recommended	Nice to have
Undertake cost-benefit analysis over facility's life cycle of implementing environmentally friendly measures	Include specific measures, demonstrated to be cost-effective	
Develop and implement a local sustainability purchasing policy	Identify environmentally friendly suppliers for key materials	
The choice of major building materials should be cognisant of each material's environmental impact, and low-impact products should be chosen where feasible	Specify all major building materials with low environmental impact/ embodied energy	Specify all major and minor building materials with low environmental impact/ embodied energy
Specify that 50 per cent of materials are manufactured/ sourced in the country where the project takes place	Specify that 20 per cent of building materials are manufactured/sourced within a radius of 200 miles	
Assessment should be made of feasibility of using recycled materials	Specify 25 per cent of recycled materials used in the main materials comprising 75 per cent by volume	Specify 50 per cent of recycled materials used in the main materials comprising 75 per cent by volume

Table 12.1: (Continued)

Minimum Standard	Recommended	Nice to have
All timber and wood products to be sourced in the country where the project takes place; all timber and wood products to be from sources certified as sustainable	Use systems furniture and task seating that is Greenguard or equivalently certified or registered; composite wood or agrifibre products should not contain urea-formaldehyde resins	25 per cent of timber and wood products to be from recycled materials
Consider salvaged, refurbished, or used furniture and furnishings (subject to HSE compliance)	All specifications to include the requirement to design out waste, e.g. prefabrication, use standard lengths	
Asbestos should not be used in any new building or refurbishment	Asbestos to be removed from all existing buildings	
Specify long-life materials based on likely usage and wear, e.g. flooring, furniture	Specify recycled (cradle-to-cradle) products, e.g. carpets, chairs etc.	
Meet or better industry limits and local VOC limits for adhesives and sealants, and for paints and coatings	Specify low-VOC products throughout the building	
Refrigerants to have no impact on ozone depletion or global warming potential	Insulating materials to have no impact on ozone depletion or global warming potential	
Provide accessible area dedicated to separation, collection and storage for dry recycling and wet waste	Provide facilities to recycle/compost 50 per cent of organic food waste on site	Provide facilities to recycle/compost 100 per cent of organic food waste on site
For construction projects, ensure that constructor sorts and recycles construction wastes	Constructor to monitor waste generated as a benchmark for future projects	Target 50 per cent of waste spoil to be reused on site
Implement a waste-management plan that recycles at least 50 per cent of construction, demolition and packaging waste	Implement a waste-management plan that recycles 75 per cent of construction, demolition and packaging waste	Specify salvaged, reused or refurbished materials for 10 per cent of building materials
Target to reuse/maintain a minimum 25 per cent of elements such as walls, floor coverings, ceiling systems by area for fit-out works	Target to reuse/maintain a minimum 50 per cent of elements such as walls, floor coverings, ceiling systems by area for fit-out works	Reuse/maintain 75 per cent of existing non-shell elements such as walls, floor coverings, ceiling systems by area

Endnotes

1 Allen, D. T. and Shonnard, D. (2001), *Green Engineering: Environmentally Conscious Design of Chemical Processes*, Prentice Hall, Englewood Cliffs, ch. 13, p. 4

2 www.wrap.org

3 www.ice.org.uk/knowledge/specialist_waste_board.asp
4 Katherine Adams presentation, Halving C,D & E waste to landfill by 2012, Strategic Forum for Construction, 15 July 2010
5 http://www.wrap.org.uk/construction/
6 http://www.wrap.org.uk/construction/tools_and_guidance/site_waste_management_planning/site_waste_1.html
7 Hardie, Marie; Khan, Shahed; Miller, Graham; of the School of Engineering, University of Western Sydney; Waste minimisation in office refurbishment projects: an Australian perspective; University of Western Sydney. Reproduced by permission of Mary Hardie

13 Water Conservation

Many buildings use much more water than they need to, leading to higher-than-necessary costs and environmental impacts. Water use can be reduced both by influencing user behaviour and by replacing older or faulty fittings. Often, measures can be taken that cost little or nothing to implement, but result in immediate bottom-line benefits.

By reducing water use, both water and sewerage bills will decrease and this could result in a considerable saving. Buildings differ in the facilities they include, but a typical consumption pattern for offices is shown in Figure 13.1, with the vast majority of use being accounted for by the washrooms[1]. The biggest single determinant of total use is likely to be the number of people working in the office. Their education in water use will, therefore, be an important issue, as well as the issue of the type of fittings installed and their maintenance.

Chapter Learning Guide

This chapter will provide an overview of the water-conservation measures for domestic and commercial developments.

- Water-saving devices to conserve and reduce water consumption that can be easily retro-fitted;

(*Continued*)

Sustainable Refurbishment, First Edition. Sunil Shah.
© 2012 Sunil Shah. Published 2012 by Blackwell Publishing Ltd.

- Checklist of water-conservation measures to be assessed against the level of refurbishment being performed.

Key messages include:

- Water efficiency can provide significant cost savings and measures can be implemented.

13.1 Performing a Water Audit

The first step in an existing operational building is to understand the current water consumption: how the water is being used; at which specific times; and which areas use the most water.

The first stage is to set up a programme to measure existing consumption and to confirm this from water bills provided. Where accurate, water bills can be used to ascertain historical consumption. The results should be profiled in tabular form to facilitate the assessment of consumption and the reviewing of trends. This should be performed for the incoming volume of water and for the volume of water discharged. Where possible, meter readings should be taken, either on a monthly basis where consumption is low, or weekly if the usage is much higher. This will provide a greater level of clarity to identify trends and highlight excessive usage and wastage from leaks.

Typically, areas of high consumption within offices and retail units are the toilet areas, with the consumption being from the sanitary ware comprising WCs, urinals and wash-hand basins. In comparison with offices and retail units, however, there is relatively low water consumption from a restaurant facility (see Figure 13.1).

The next stage is to identify where water is being consumed within the building and to reconcile that with the anticipated or predicted volumes based on meter

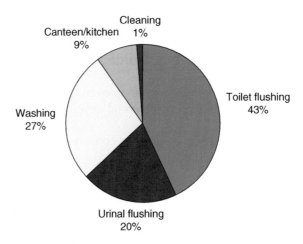

Figure 13.1: Water usage in a typical office building.

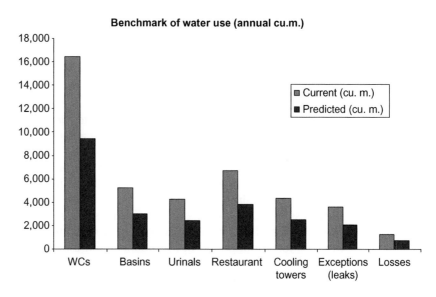

Figure 13.2: Typical consumption vs. predicted consumption.

Table 13.1: Template for the collection and calculation of water data (adapted from Environment Agency – Waterwise)

Item	Number of units [A]	Low rate (L/flush or L/min) [B]	Operating time (mins/day or uses per day) [C]	Water used (l/day) [A × B × C]	Comments
WCs	30	9 L/flush	5 flushes/day	1,350	
Urinals	10	7.5 L/flush	7 flushes/day	525	Office hours only
Taps	50	4 L/min	3 mins/day	600	
Showers	03	10 L/min	10 mins/day	300	

size, sub-metering or typical industry figures. A graph can be generated to represent the findings (see Figure 13.2), which highlights clearly where the greatest focus should be made. In this example, excessive usage can be seen in all areas, with a particularly high consumption in the toilets and the restaurant. By concentrating on reductions in these areas, the largest savings can be made with the minimum effort.

Much of the data required will need to be estimated, based on knowledge of the units within the building. An example is shown, in Table 13.1, of how to calculate the water consumption from toilets, urinals, taps and showers – a similar methodology should be used for all water appliances, including those in the restaurant. For each individual appliance, identify: how many there are; the water consumption of each; and, roughly, the amount of time each appliance is in use for. For toilets and urinals, this will be based on an assumption of the flush rates and number of people within the building. By multiplying these figures, a rough estimate of water consumption can be calculated. A total of 80 per cent, or more, of the water consumed should be identified and confirmed through reconciliation with the water bills and metering within the building.

Once the highest-consuming activities and those with the greatest potential for reduction have been identified, there are a number of ways to reduce water consumption:

- Detecting leaks in supply pipes should be performed either by the local water company or by a specialist contractor. Leaks can account for up to 20 per cent of consumption and lead to structural problems if undetected;
- The maximum flush volume on toilets can be reduced to 4 litres for newly installed toilets due to changes in bowl dynamics, although older models should be maintained at 7.5 litres;
- Waterless or reduced-flow urinals offer the same service as traditional urinals when used and maintained correctly. It is recommended that these units be installed for a trial period in accordance with the manufacturer's instructions – waterless urinals are in operation at airports and have posed no difficulty;
- A variety of mechanisms are available for the collection and storage of rainwater for subsequent use on site, with a number of proprietary systems available to provide a complete system; the water can be used primarily for toilets;
- The use of 'grey water' is common in Japan and Germany where it has been in use for the past 20 years.

Typical water-consumption figures for an office environment are in the region of 15 to 20 m^3 per FTE (full time equivalent), inclusive of the use by visitors, based on previous experience. Best-practice figures of below 8 m^3 per FTE have been set, with the UK Environment Agency having achieved 7.7 m^3 through the efficient use of 'grey water' and the education of staff. Benchmarks should be set, based on realistic targets and a project plan. A site using 20 m^3 per FTE could look to set a 10 per cent year-on-year reduction target over a three-year timeframe.

A water-management plan can be developed, classifying the no-cost, low-cost, and medium-to-high cost options, with their respective payback periods. This will provide a priority list for the implementation of the various water-management projects and highlight where the greatest workload is going to be. In addition, it will be possible to provide a normalising factor of carbon dioxide from which to assess performance-measurement data.

13.2 Reducing Water Use

Often, water can be wasted simply because staff are not aware of the concept of using too much water. A staff-awareness campaign, highlighting the amount of water used in particular areas, can be very effective. It can also be extremely effective to ask staff for suggestions on saving water. Because they use the equipment every day, they will have the best understanding of where water can be saved. Staff might also know about specific equipment that is wasting water due to a need for maintenance (such as scaled-up taps), or due to poor installation (such as long, uninsulated pipe runs, leading to a long wait for hot water).

Toilet-Water Use

A significant proportion of water consumed in offices is used in toilets. The amount of water that the toilets will use depends on their age – if they are more than 10 years' old, they probably use 9 litres for each flush. Even if they are only five years' old, they can use up to 7.5 litres. Modern toilets have a 4-litre flush, and, if having a dual-flush cistern, an even smaller flush of only 2 litres. If you are refurbishing your washrooms, dual-flush toilets that offer users a choice between a full flush or a half-flush can be effective; they are available for toilets with either visible or concealed cisterns.

If the toilets are not being replaced, consider reducing the flush artificially by installing a cistern-displacement device (such as a 'Hippo' or 'Saveaflush') or, possibly, a retro-fit, variable-flush device. This will result in a reduced flush, however, it is important to remove them if users find that they do not provide sufficient water volume – otherwise, their installation might result in even more water being used. Many water companies provide such devices at a reduced rate or even free of charge.

In all cases, it is important that you check regularly that toilet cisterns and cistern devices are working properly, and not leaking. In addition, making a regular check that the cistern-fill level is set to the correct mark will ensure that excessive water is not used, and that the cistern operates properly.

Taps and Showers

Washing accounts for 27 per cent of water use in typical offices, so that choosing the right fittings can have a considerable effect on water bills. Taps that can be left on by users, or those dripping, might be wasting many litres a day. Even low-water use taps, such as spray taps or push-top taps, can lose water if they are poorly maintained, especially in hard-water areas where they can easily scale up. Many tap controls can be fitted to existing taps, provided that they are of the 'pillar' type, and cost very little to install.

Although you might have seen water-conservation advice that recommends installing showers rather than baths, a power shower can fill the equivalent of a bath in only a few minutes – consider whether a power shower is really needed, or whether a gravity or lower-power version would be equally good.

When a planned refurbishment of the office washroom is due, the opportunity arises to select taps and showers that conserve water. Showers with a restricted water usage, such as 9 litres per minute, can be specified. These will still provide a good flow of water with reduced wastage. For hand basins, a tap that has a restricted outlet (such as that of aerated taps) with a good wash pattern will be effective. Such taps will use much less water than will an open-ended tap. There are many different types of tap mechanism available, and some have other benefits, such as reduced risk of scalding (from thermostatic, mixer taps) and health-and-safety benefits (from infra-red taps that operate without being touched by users).

Canteens and Kitchens

Even if your office includes only tea points, rather than a canteen, it could still be using a lot of water. Most of the water used in such areas is attributable to the users' behaviour, such as filling sinks, rather than washing up under a running tap. Make sure that all users realise how much water they could save and that they are aware of ways to help. The equipment within a kitchen will influence how much water is used. If the kitchen is equipped with a dishwasher, check the water consumption of the model and the most efficient setting (this is usually included in the instructions that accompany the machine). Most people use only one setting, so it makes sense to check that it is the most effective one!

Consider how water is used in the kitchen. If it is simply for rinsing cups and spoons, the main water use will be the running of taps. You could help users to reduce water by the use of stickers (usually available from your water company) or by changing the taps to spray or push-top type. However, these types of tap are not really suitable if you need to run a full sink, so this does need to be first discussed with the users. If you are specifying a new dishwasher, check the energy label – this also tells you how much water the machine will use on the 'eco' setting. Dishwashers designed for eight place settings will use between 10 and 30 litres of water per wash, whereas full-size (12 place settings) machines will use 10–50 litres. The more economical machines actually use less water than the amount required for filling two sinks for washing up; if the office does not have a dishwasher, it might be worth purchasing one. In commercial kitchens or canteens, the equipment used will be very different. In these situations it will be easier to train the users to adopt water-conservation measures, since there will be a limited number of people with access to the equipment. Commercial dishwashers and glass-cleaning equipment can be specified to minimise the volume of water used, and some machines even reuse the last (clean) rinse water in the next pre-wash. Tell your commercial supplier that you are interested in products that save water and ask what is available.

Grounds' Maintenance

Many offices have little open space other than car-parking areas, thus avoiding the need to use water to maintain the grounds. It is usual to have a few plants around the reception area within the office, or interspersed between parking bays in the car park; some offices have grounds with large areas of 'green' landscaping that need maintenance. The most important issue for grounds' maintenance is to reduce, or eliminate, the watering of plants. It is possible to eliminate watering altogether where the grounds and the planting are planned carefully, however, if watering is considered to be essential, an efficient watering system should be used. It is also possible to store rainwater for watering to reduce the amount of mains water that is used.

13.3 Rainwater Harvesting

Rainwater harvesting is common for toilets; new technologies allow for the reuse of water in toilets from the recycling of treated sewage effluent. When this is combined with the use of water-efficient appliances and behavioural changes, demand can be significantly reduced.

The mains water supply undergoes a complex and energy-consuming cycle that includes its treatment to be fit for drinking and its pumping from the reservoir to our homes and places of work. Hence, the opportunity to utilise technologies that save water should be considered as part of any major refurbishment.

Figure 13.3: A conceptual rainwater harvesting system (source CIRIA SUDS Manual C697).

At its simplest level, a rainwater-harvesting system could be simply positioning a water butt to collect rainwater run-off for subsequent reuse in garden and land-scaping areas. More sophisticated systems (see Figure 13.3) are now commer-cially available that can be used to collect rainwater run-off from larger roof areas, paved and hard-landscaping areas, then stored, filtered and subsequently pumped to provide water for non-potable use within the building (usually limited to flush-ing WC cisterns).

Rainwater-harvesting systems should not be confused with 'grey-water' systems. 'Grey-water' systems are designed to reuse the waste water from ablutions for other non-potable purposes, usually flushing WCs.

The main criterion affecting a rainwater-harvesting system is the size of the water tank, which needs to be sufficient to provide adequate storage capacity. The storage requirement will be determined by a number of interrelated factors. They include:

• Local rainfall data and weather patterns;
• Roof (or other) collection area;
• Run-off coefficient (this varies according to roof material and slope);
• User numbers and consumption rates.

In low-rainfall areas where the rainfall is of uneven distribution, more care has to be taken to size the storage properly. During some months of the year there might be a surplus of water, whilst at other times there could be a deficit. If there is insufficient water throughout the year to meet the demand, then sufficient storage will be required to bridge the periods of scarcity. As storage is expensive, this needs to be achieved carefully to avoid unnecessary expense.

On days when rainfall is heavy, the flow into a tank is higher than the outflow drawn by water users. A small tank will soon become full and then start to overflow. An **inefficient** system is one where, taken over, say, a year, that overflow constitutes a significant proportion of the water flowing into the tank. Insufficient storage volume is, however, not the only cause of inefficiency: inadequate guttering will fail to catch water during periods of intense rain; leaking tanks will lose water; and an 'oversize' roof will intercept more rainfall than is needed.

Before the retro-fitting of a rainwater-harvesting system into an existing building, a number of issues need to be addressed. Not least of these is the ability to accommodate a storage tank (typically 12,000–25,000-litre capacity), either below ground or at a position in the building to which rainwater collection can easily be diverted (such as space in a basement).

If tanks have to be buried below ground, this will incur major installation costs. Similarly, additional costs might be incurred because the water collected needs to be separated from the water for drinking, usually being used only for flushing toilets; modifications to internal plumbing installations will also be required.

13.4   Flood Risk and Sustainable Drainage Systems

Sustainable drainage systems (SUDS) should be considered for major refurbishments to reduce the potential risk from surface-water flooding. Assessments of flood risk should be undertaken to provide a technical assessment of all forms of flood risk to the development and its surrounding area. Where there are potential risks associated with surface-water drainage and/or flooding from sewers, they should be addressed. Properties can be refurbished to be resistant to flooding by the inclusion of flood-resistant materials.

Some of the measures that could be adopted are:

- Examining evidence of sub-soil porosity and suitability for use of infiltration SUDS;
- Undertaking pre-development and estimated, post-development run-off calculations to determine the scale of SUDS required;
- Assessing flood risk where this is deemed appropriate;
- Making proposals for integrating the drainage system into the landscaping or into public, open space;
- Demonstrating good ecological practice, including habitat enhancement;
- Estimating land take-up for different drainage options, based on initial calculations from any significant drainage structures.

In many cases, where refurbishments do not change the footprint of the building, applicable to many of the minor and medium-scale projects, there is likely to be no change in surface-water run-off. Therefore, the potential for increased flood risk is limited. It is worth being aware of the variations in rainfall in the area and any potential flooding events further upstream.

13.5   Water-Conservation Checklist

Table 13.2 provides a summary of the items to be discussed during a sustainable-refurbishment project, capturing where specific items are necessary, recommended or nice-to-have options.

Table 13.2: Water-conservation checklist

Minimum Standard	Recommended	Nice to have
Perform a water audit to confirm current usage of water and which mechanisms to reduce	Set target to reduce water use by 25%	Set target to reduce water use by 40%
Maximise water efficiency to best-practice-in-country guidelines or to 10 m³ per workstation	Exceed the potable-water-use reduction by an additional 20%	Exceed the potable-water-use reduction by an additional 50%
UV filtered water provided for drinking water (not bottled)		
Attenuate site-water run-off to natural and municipal watercourses (e.g. porous pavements) and to protect against water-borne contamination	Manage peak-flow site-water run-off not to exceed pre-construction levels	
Install low-flow fixtures and fittings		
Point-of-source, electrically heated hot water in coffee areas and toilets		
Rainwater harvesting to maximise rainfall	Consider the use of sustainable drainage systems to utilise run-off and 'grey water'	Maximum utilisation of 'grey water'
Install sub-metering equipment to statutory requirements	Sub-metering of all major water-using plant; install water-leak detection; quarterly monitoring of water use	Smart metering with remote data loading; install water-leak detection; annual targets for water use reduction
Install water-efficiency measures (e.g. dual-flush, 3/6 ltr toilets, percussion/aerated taps, waterless urinals)	Install leak-detection system to identify all major leaks, internal and external to the building, covering all mains-water supplies to the building	WC wastes to an anaerobic digester for composting and biogas generation
Recommend not to build if the lowest point of the development is below a defined flood level; undertake a flood-risk assessment as appropriate	Design in flood mitigation if development is in a 1:50-year flood zone	Design in flood mitigation if development is in a 1:100-year flood zone
Landscape design to incorporate local flora and fauna capable of surviving without watering	Drought-resistant landscape design	

Endnote

1 Key Performance Indicators for water use in offices, CIRIA W11, February 2006

14 Biodiversity

The refurbishment of buildings has the potential to improve the ecological value of a site, with initiatives such as green roofs, living walls and green spaces all being able to improve biodiversity, as well as making workers happier and more productive[1]. This provides a mechanism to increase positive impacts – for people, wildlife and the economy. Refurbishments can create habitats in which wild species thrive, and something that we can all enjoy.

Chapter Learning Guide

This chapter will provide an overview of including biodiversity measures into the refurbishment of buildings.

- Inclusion of wildlife considerations and additional green infrastructure within a development;
- Benefits of green roofs and walls, various approaches and challenges to their development;
- Checklist of minimum requirements and good practice to implement sustainable-material refurbishments.

Key messages include:

- There are a variety of types of green roof to apply to many types of roof and walls ;
- Biodiversity not only supports wildlife, but also improves staff well-being.

Sustainable Refurbishment, First Edition. Sunil Shah.
© 2012 Sunil Shah. Published 2012 by Blackwell Publishing Ltd.

14.1 Introduction

Detailed design of buildings and other structures should include specific measures for biodiversity. Such features include nesting or roosting sites and structures, 'living roofs and facades', building-integrated vegetation and trees into street design and 'hard' open spaces[2]. The amount of green spaces and nest sites that should be incorporated into refurbished developments should be guided by what is locally appropriate, and the relevant advice on the amount of provision, its location within the development, its siting and all associated information should be sourced from an experienced ecologist.

Case study – Horniman Museum extension living roof[3]

The extension to the Horniman Museum, in Forest Hill, south London, included building the new Centre for Understanding the Environment (CUE), incorporating a 400-square-metre and a 250-square-metre, pitched, extensive green roof (see Figure 14.1). The extension living roof was designed with the conservation of biodiversity as an objective from the beginning. The roof provides more biodiversity than a sedum roof, as well as providing a sound barrier, increased insulation, rainfall attenuation, summer cooling and sustainable drainage. The roof also has an educational benefit as a point of great interest to museum visitors.

A ten-year survey found that the roof has developed into species-rich grassland, supporting a number of plants notable to London. The south-facing section is sandy and dry, dominated by grasses. The roof supports abundant meadow wildflowers and taller meadow grasses on the wetter north-facing section, and gaps in the turf have allowed further plant species and mosses to flourish. Upkeep of the roof has required minimal intervention, and it continues to prosper, requiring only occasional watering and annual mowing. However, the original intention to cool the building structure in summer, through increased evapo-transpiration by way of an automatic irrigation system, has been abandoned owing to the system clogging with algae. This problem could be avoided in the future by using modern design approaches.

Green infrastructure networks can have a number of benefits for refurbished developments. In June 2009, Copenhagen introduced a mandatory requirement for green roofs, joining Portland in the US, which has been a front runner in the introduction of green roofs, principally for storm-water management. The following benefits are provided:

- Reduction in temperatures; on hot, sunny days, rooftop temperatures may be up to 40°C cooler with a green roof than with a conventional flat, dark-coloured roof[4];
- Reduction of rainwater run-off by absorbing and slowly releasing large amounts of water;

Figure 14.1: Green roof of the CUE building at the Horniman Museum.

- Reduction of noise for occupants, especially on upper floors;
- Increased urban biodiversity by providing habitat space for birds and small animals;
- Increased evaporative cooling effect by the retention of storm water in vegetation.

Street trees, and trees in public spaces (especially large, broad-leaved trees), help to alleviate the effects of climate change. For example, trees provide shade in the summer, reducing the need for mechanical air-conditioning (depending on their proximity to buildings). Trees also provide natural cooling systems as they consume large amounts of available energy in the atmosphere through the process of evapo-transpiration. It is important to ensure that the right trees are planted to cope with present and future local conditions.

- Trees (when large enough) provide shade – surface, peak-temperature reductions of between 5°C and 20°C may be possible;
- Evapo-transpiration through trees and vegetation can result in the reduction of peak summer temperatures by between 1°C and 5°C.

There should be well considered and deliverable proposals for the long-term management of green space and for funding to ensure such management as is required for maintaining and bedding in. This will be supported by a biodiversity action plan, identifying the key species and presenting strategies designed to improve and increase habitat for selected species and increase their numbers.

14.2 Green Roofs and Walls

Green roofs are vegetated layers that sit on top of the conventional waterproofed roof surfaces of a building. Whilst green roofs come in many different forms and types, usually a distinction is made between extensive, intensive and biodiverse (or wildlife) roofs. These terms refer to the degree of maintenance the roofs require.

Intensive green roofs are composed of relatively deep substrates (20cm+) and can therefore support a wide range of plant types: trees and shrubs as well as perennials, grasses and annuals. As a result, they are generally heavy and require specific support from the building. Because of their larger plant material and horticultural diversity, intensive green roofs can require a substantial input of resources – the usual pruning, clipping, watering and weeding as well as irrigation and fertilisation.

Extensive green roofs are composed of lightweight layers of free-draining material that support low-growing, hardy, drought-tolerant vegetation. Generally, the depth of growing medium is from a few centimetres up to a maximum of around 10–15cm. These roof types have great potential for wide application because, being lightweight, they require little or no additional structural support from the building. Furthermore, because the vegetation is adapted to the extreme rooftop environment (high winds, hot sun, drought, and winter cold), extensive green roofs require little in the way of maintenance and resource input. Extensive green roofs are typically favoured for refurbishments.

Biodiverse (or wildlife) roofs are becoming more popular, and are designed to replicate the specific habitat needs of a single, or small number of, species, or to create a range of habitats that can maximise the array of species that inhabit and use the roof. Creating a wildlife roof could enable some demolition waste to be placed onto the roof. Care would need to be taken to screen the waste to remove contaminants and any seed banks mixed in it, which might lead to the resulting green roof requiring an unexpectedly high level of maintenance.

Benefits of a Green Roof

Green roofs have numerous social, economic and environmental benefits and can contribute positively to issues surrounding climate change, flooding, biodiversity and declining green space in urban areas.

Reducing storm-water run-off as part of a sustainable-drainage-system (SuDS) strategy to support the rainfall changes due to climate change, SuDS filter, absorb and moderate flows of run-off. SuDS help to reduce pollution of watercourses and localised flooding as well as providing amenity and biodiversity benefits. Green roofs are one method of controlling storm water at source (i.e. closest to the source of the precipitation) under a SuDS strategy. Green roofs are much easier to retro-fit in the urban environment than are many other SuDS components, so their potential for reducing storm-water problems is significant.

Once established, a green roof can significantly reduce both peak-flow rates and total run-off volume of rainwater from the roof, compared to a conventional roof. Green roofs store rainwater in the plants and substrate, and release water back into the atmosphere through evapo-transpiration. In Germany, it is recognised that a green roof will have a positive effect on storm-water run-off.

Increasing the roof's lifespan due to reducing the impact of thermal fluctuations on the roof. The original green roofs in Germany were created in the 1880s when it was typical to cover bitumen with 6cm of sand to protect the bitumen from fire. The sand was also found to extend the life of the waterproof layer and was integrated naturally with the vegetation. Green roofs have now been shown to double the lifetime of the waterproofing membrane below the green roof by creating a barrier that protects the waterproofing membrane from harm.

Reducing energy use by improving the roof's thermal performance, although the level of improvement depends on the daily and seasonal weather conditions. By retro-fitting green roofs, both air-conditioning and heating usage is decreased. During the summer, solar energy is utilised by plants for evapo-transpiration, reducing the temperature of the green roof and the surrounding microclimate. During the winter months, a green roof can add to the insulating qualities of the roof. However, thermal performance is extremely dependent on the amount of water held within the green roof's substrate. Water has a negative effect on thermal conductivity. Thus, in a damp, winter climate, a green roof will add little to the overall thermal performance of the roof. Green roofs are not assigned a fixed U-value as they are assumed to hold water.

Urban Heat Island Effect is managed on existing buildings through 'future proofing'. Green roofs are one of the most effective ways of combating the urban-heat-island effect and will therefore be part of the raft of future measures designed to help cities adapt. The urban-heat-island effect is the temperature disparity between urbanised areas and surrounding rural areas. Urban landscapes have a much higher proportion of dense, dark, impermeable surfaces which have a low albedo (reflectivity). This means they absorb heat, unlike plants, which reflect it. This stored heat is re-radiated at night, thus warming the city more than the surrounding countryside. This can make city centres up to 7°C warmer than the surrounding countryside due to the urban-heat-island effect[5].

Increasing biodiversity and wildlife according to the type of green roof, and type of vegetation and substrate they contain. Roofs designed to replicate the habitat either for a single, or for a limited number of, plant or animal species are often referred to as biodiverse roofs. They can be especially important as a tool to recreate the pioneer (wasteland) communities that are sometimes lost to redevelopment. It is often the neglected brownfield sites in urban areas that are the most biodiverse. The best biodiverse roofs support a range of habitats for wildlife through a range of substrates, depths and micro-habitats.

Reduced sound transfer from rain hitting the hard roof is achieved as the sound is absorbed by the combination of the growing medium, plants and trapped layers of air within the green roof system; the green roof acts as a sound-insulation barrier. Green roofs have been employed successfully as a means of sound abatement along new runway approaches at Frankfurt International airport and Schiphol airport in Amsterdam.

Amenity space, adding value to buildings with improved views, making buildings easier to let. Accessible roofs, designed to allow people to relax, attend events or participate in gardening, can make a real difference to the way in which people use and enjoy buildings.

Table 14.1[6] highlights the benefits, as a relative comparison between each of the roof types, and demonstrates the benefits of the deeper-substrate roof types, which are able to cater for greater biodiversity species and improve thermal performance. This will need to be tempered against the loading capacity of the existing structure as the deeper-substrate roofs are much heavier (see Table 14.2), so that both issues can be effectively managed.

Table 14.1: Relative benefits of the different green roof types

Roof Type	Potential Benefit					
	Climate Change	Energy	Urban Heat	SuDS	Biodiversity	Amenity
Intensive	✓✓	✓✓	✓✓✓	✓✓✓	✓	✓✓✓
Extensive (<40mm)	✓	✓	✓	✓	✓	✓
Extensive (>40mm)	✓✓	✓✓	✓✓	✓✓	✓✓✓	✓✓
Recreation	✓	✓	–	–	–	✓✓✓

Table 14.2: Indicative structural loading for various types of roof (when fully saturated)

Roof Type	Loading (kg/m^2)
Gravel surface	90–150
Paving slabs	160–220
Vehicle	From 550
Extensive green roof (sedum mat)	60–90
Extensive green roof (substrate)	80–150
Intensive green roof	200–500

Chicago City Hall

Project Name: Chicago City Hall
Year: 2001
Cost: US$2.5m
Owner: City of Chicago, Dept of Environment
Type: Semi-Extensive, Test/Research
Size: 20,300 sq.ft.
Slope: 1.5 per cent
Access: Accessible, Private

Figure 14.2

(*Continued*)

As part of an EPA study and initiative to combat the urban-heat-island effect and to improve urban air quality, Mayor Richard M. Daley and the City of Chicago began construction of a semi-extensive green roof in April 2000. It was completed in the summer of 2001 at a cost US$2.5 million, and serves as a demonstration project and test green roof. It is monitored for plant survival as well as other environmental features. Chicago City Hall's green roof saves US$5,000 a year on utility bills.

Figure 14.3

Chicago's most famous rooftop garden sits atop City Hall, an 11-storey office building in the Loop. City Hall and the adjacent Cook County building appear to most people as one building, spanning a city block bounded by LaSalle, Randolph, Clark and Washington streets. First planted in 2000, the City Hall rooftop garden was conceived as a demonstration project – part of the City's Urban Heat Island Initiative – to test the benefits of green roofs and how they affect temperature and air quality. The garden consists of 20,000 plants of more than 150 species, including shrubs, vines and two trees. The plants were selected for their ability to thrive in the conditions on the roof, which is exposed to the sun and can be windy and arid. Most are prairie plants native to the Chicago region.

Like all green roofs, the City Hall rooftop garden improves air quality, conserves energy, reduces storm-water run-off and helps lessen the urban-heat-island effect. The garden's plants reflect heat, provide shade and help cool the surrounding air through evapo-transpiration, which occurs when plants secrete or 'transpire' water through pores in their leaves. The water draws heat as it evaporates, cooling the air in the process. Plants also filter the air, which improves air quality by using excess carbon dioxide to produce oxygen.

The rooftop garden mitigates the urban-heat-island effect by replacing what was a ballasted, black, tar roof with green plants. The garden absorbs less heat from the sun than did the tar roof, keeping City Hall cooler in summer and requiring less energy for air-conditioning. The garden also absorbs and uses rain water. It can retain 75 per cent of a 1-inch rainfall before there is storm-water run-off into the sewers.

Data Comparison Between the City Hall Rooftop Garden and the Black Tar Roof of the Cook County Building

As part of its Urban Heat Island reduction plan, the City will work with scientists from the US EPA and Lawrence Berkeley National Laboratory to estimate the effects of gardens such as this on a neighbourhood or a city. Weather stations were placed on both the City Hall and County sides of the building to compare air temperature and other data between the two rooftops. In addition, an infra-red thermometer was used to measure surface temperatures.

On 9 August 2001, at 1:45pm, when the temperature was in the 90s, the following surface temperature measurements were obtained:

City Hall Roof (paved):   126 –130°F
City Hall Roof (planted): 91–119°F
County Roof (black tar):   169°F

That's at least a 50°F difference between the garden roof and a black roof!

Design of Chicago's City Hall Rooftop Garden

The design of the Chicago City Hall rooftop garden required many months of work by environmental engineers, structural engineers, architects and landscape architects. The plan for the rooftop garden was designed so that any of the standard green roof systems could be adapted to this project. Structural considerations, water storage and irrigation, and green-roof-system layers were specified in the design.

In its design, the City Hall rooftop-garden design team attempted to maximise the square footage of roof that could sustain a green roof system whilst maximising the amount of space available for an intensive system. The limiting factor for this type of system is the support structure of the existing roof. The City Hall rooftop-garden design indicates three types of system – intensive, 'semi-intensive' (the shallow end of an intensive system), and extensive.

The existing conditions and use of the rooftop were addressed by the design, including the need for maintenance access. Also, it was determined that the framing of former skylights would remain and would need to be incorporated into the design. Even though the rooftop would not be

(Continued)

Figure 14.4

accessible to either the general public or the building occupants, the demonstration nature of the project made it desirable to incorporate a variety of features in order to create a diversity of green roof venues.

Construction of the Chicago City Hall Rooftop Garden

Typically, a green roof begins with an insulation layer, a waterproof membrane to protect the building from leaks, and a root barrier to prevent roots from penetrating the waterproof membrane. A drainage layer, usually made of lightweight gravel, clay, or plastic, comes next. The drainage layer keeps the growing media aerated in addition to taking care of excess water. Since a green roof system covers the entire roof, drainage points must be accessible from above for maintenance purposes. On top of the drainage layer, a geotextile, or filter mat, allows water to soak through but prevents erosion of the fine soil particles. Finally, the top layers consist of growing media, plants, and a wind blanket. The growing media are lightweight material that helps with drainage whilst providing nutrients to the plants. A wind blanket is used to keep the growing media in place until the roots of the plants take hold.

http://www.cityofchicago.org/city/en/depts/doe/supp_info/green_roof_systemslayers.html
   http://www.greenroofs.com/projects/pview.php?id=21

Challenges and constraints

There are also a number of issues that need to be carefully considered prior to the installation of a green roof to ensure that it is appropriate and is also the best form of green infrastructure and biodiversity that can be incorporated into the site.

Maintenance All roofs need to be maintained. Extensive green roofs require an annual inspection to ensure that all drainage outlets and shingle perimeters are vegetation-free, with any vegetation requiring only an additional low level of maintenance, depending on the system. Intensive and semi-intensive green roofs will need irrigation and a higher level of maintenance, and should be treated as a garden.

Structure Green roofs will mean an additional strain on the structure to provide support. Regular inspections should be made regarding the integrity of the build-ing to ensure that the additional loading is safe. For refurbished buildings, it is common to provide an extensive roof, due to the reduced weight of the green roof. However, where sufficient roof loadings are available, such as those that result from the removal of heavy plant, intensive roofs can be installed.

Cost Green roofs cost more to install than a traditional roof, given the additional labour and materials required for a typical roof, regardless of the type of roof. Costs can be offset principally through energy savings, and also where flood mitigation measures are not required. Additional intangible benefits, such as employee well-being and corporate responsibility can also be captured. There is no direct relationship between green roofs and asset value that has been identi-fied, but a green roof can help to achieve a BREEAM and LEED rating, which, in turn, can help to secure a higher asset value.

A report, published in 2003 by English Nature, provides an overview of the average costs of a green roof in a number of countries. The costs are for extensive, sedum, matted, green roofs[7]. The variation in costs is generally attributable to the size of the roof: the larger the roof, the lower the costs.

- USA: US$150–200/m²
- Germany: €20–40/m²
- UK: £85–93/m²

Fire Green roofs are seen to be more of a fire risk than more conventional roof solutions. However, in Germany, a building with a green roof will be eligible for a reduction in the cost of fire insurance, as the green roof protects the waterproof-ing membrane from fire. Construction needs to follow strict rules regarding the amount of combustible material and the need for firebreaks. As a result, there have been no major incidents of fire in the German and Swiss markets where over seventy-five-million square metres of green roofs have been installed.

Damage to Waterproofing Green roofs by themselves do not cause roofs to leak. Typically, this is as a result of poor installation and a lack of rigorous leak testing before installing the green element. There is evidence from Germany that green roofs protect the membrane from damage and therefore reduce the potential for leaks. It is, however, critical to ensure that the design, installation and testing are performed by an experienced organisation.

Green Walls

Green walls are essentially a living, and therefore self-regenerating, cladding system using climbing plants. Whilst climbers have been used traditionally on buildings for centuries, contemporary architecture using high-tensile steel cables have enabled the concept to be used far more adventurously. With suitable species' selection, heights of up to 25 metres can be attained.

The most commonly used species for wall-greening are ivy (*Hedera* sp.), Russian vine (*Fallopia* sp.) and Virginia creeper (*Parthenocissus* sp.), which can climb directly onto wall faces, especially those of brick and stone where the porous surface allows them to attach more easily, albeit the installation of trellises and wires can aid their growth. All can relatively quickly form a dense, evergreen foliage many metres in height, and on many older buildings this is actively managed to ensure that it does not obscure windows and other openings. Such dense foliage provides an effective nesting habitat for a variety of birds, including robin, wren and blackbird, as well as serving to baffle noise. A number of varieties have colourful foliage that will change with the seasons, often turning red or golden during autumn.

Although it is often thought that these climbing plants can damage wall surfaces through their tendrils and rootlets (in the case of ivy), in most cases, provided that the wall is solid and well built, there is no reason why damage should occur. Care should be taken, however, in respect of walls with cavities or with crumbling mortar in which roots could take hold and expand.

Large-scale use of green walls is still a new concept, however, there is a solid research base, which has developed in Germany over the last 20 years. Active research is a major focus for several German academic institutions[8]. The new approach, being pioneered in Switzerland and Germany, is to see them as integral to the design concept. Clearly, this makes sense both in visual terms and in designing-in practical features.

Green walls have a number of direct and indirect benefits, including:

- Climbers can provide significant levels of shading to the walls from the sun, with daily temperature fluctuation being reduced by as much as 50 per cent. The effectiveness of this cooling effect is related primarily to the total area shaded, rather than to the thickness of the climber. The use of climbers to reduce solar heating is most effective if they are used on the wall that faces the sun, and also on the west wall, which experiences afternoon heating.
- Evergreen climbers provide winter insulation, not only by maintaining a pillow of air between the plant and the wall, but also by reducing wind-chill on the wall surface. The effectiveness of winter insulation is related to the thickness of growth, which is generally related to the age of the plant. In some cases, however, growth patterns change as the plant ages, for example, there might be a reduction in the dense, twiggy growth that forms the most effective insulation.
- Climbers on buildings can help protect the surface of the building from damage from very heavy rainfall and hail, and possibly can play some role in intercepting and temporarily holding water during rainstorms, in the way that green roofs do. They also help to shield the surface from ultraviolet light, which might be an important consideration for certain modern cladding materials.

- Reduction of solar heating of the sides of buildings helps to reduce the 'heat-island effect', whilst also absorbing carbon dioxide emissions, helping to provide a carbon sink.
- The use of climbing plants can contribute to local and regional biodiversity targets. Because they make use of vertical space, they can add considerably to the area that is potential habitat without taking up room on the ground.
- The green walls also provide an opportunity for wildlife. Any climbing plant will offer habitat to invertebrates, such as insects and spiders, which, in turn, will be food for insect-eating birds and bats.

14.3 Provision for Birds

Birds play an important role in adding to biodiversity within urban centres, with the refurbishment of buildings providing an opportunity to introduce species of birds that use buildings as a place to breed. Birds, such as the house martin, house sparrow and swift, breed extensively in buildings.

Modern buildings are designed to be weathertight with clean edges, which can exclude birds. Refurbishments, at a minor to a major scale, can incorporate bird-friendly designs into the redesigned space. Provision for birds can be designed-in very easily through creating internal nesting opportunities.

If there are difficulties in providing an internal nest, a nest box can be placed on the outside, positioned out of the direct sun, wind and rain. The location can be either on the roof, or on the side of the building. Species-specific hole sizes are important:

- 32mm for house sparrows;
- 45mm for starlings;
- swifts require a 'letter box' entrance of minimum 65 mm × 25–35 mm.

Conversely, there is a need to ensure that buildings do not pose any impact on birds, and, in situations where a high incidence of bird collisions do take place, alterations or retro-fits might be required. Solutions revolve around refurbishing problematic windows and glass facades to reduce bird collisions:

- Consider installing transparent or perforated, patterned, non-reflective window films that make glass visible to birds (examples include Scotchprint, or CollideEscape).
- Consider painting, etching, or temporarily coating collision-prone windows to make them visible to birds.
- Install louvres, awnings, sunshades, light shelves or other shading/shielding devices, positioned at large expanses of glass, to reduce reflection and to signal the existence of a barrier.
- Install and operate reflective blinds, shades or curtains to reduce glazing reflectivity and indicate the presence of a barrier to flight. Close curtains or blinds during the evenings if the interior is illuminated.
- Consider reglazing existing windows that experience high rates of bird collisions with low- reflectivity, etched, frosted, or fritted glass. Also, consider replacing large existing windows with multiple smaller units, divided lights or opaque sections.

Undertake strategies to create a physical barrier to the glass, or move features that are attractive to bird populations:

- Install exterior coverings, nettings, insect screens, lattice-work, artwork, shading or shielding devices at notably hazardous windows to deter birds, or otherwise reduce the momentum of their impact.
- Consider planting trees and shrubs close to the building within a maximum of three feet from a problematic façade or curtain wall. This planting strategy will block habitat reflections, and birds alighting in these trees will not have the distance to build momentum if they move towards the glass. This planting strategy also provides beneficial summertime shading and reduces cooling loads.
- Relocate interior plantings, water sources or other features that are causing birds to crash into glass windows.

14.4   Biodiversity Checklist

Table 14.3 provides a summary of the items to be discussed during a sustainable-refurbishment project, capturing where specific items are necessary, recommended or nice-to-have options.

Table 14.3: Biodiversity checklist

Minimum Standard	Recommended	Nice to have
Protect flora and fauna during construction (e.g. trees, hedges, watercourses, etc.)	Determine existing roof and outdoor spaces that can incorporate mitigation – green roof	
Develop pollution incident-control plan		
Use indigenous vegetation suitable for the local maintenance requirements	Utilise vegetation to attenuate rainwater	Gravel roads/parking, not tarmac (water retention)
Determine the effect of the development on existing habitats; provide mitigation measures for adversely affected areas	Encourage use of trees and planting to reduce noise intrusion to the site and local community	
Develop a local Biodiversity Action Plan (BAP) to include maintenance of the landscape		
Identify important local species and plan for their maintenance	Provide new habitats for important local species – develop as a feature (e.g. fountain or central garden) Maintain trees and planting to reduce noise intrusion to the site and local community	Consider local sponsorship to create biodiverse environments close to office (e.g. sponsor planting on a roundabout)
Ensure environmentally friendly maintenance (e.g. mowing) regimes	Maintain habitats for important local species	

Endnotes

1 UK Green Building Council study, *Biodiversity and the built environment.*
2 Environment Agency's 'Green Roof Toolkit', at http://www.environment-agency.gov.uk/business/sectors/91967.aspx
3 *Source: Biodiversity and the Built Environment: Case Studies,* Biodiversity Task Group, UK Green Building Council, 2009 http://www.ukgbc.org/site/info-centre/display-category?id=118
4 London's Urban Heat Island – a summary for decision makers, Mayor of London, 2006
5 USEPA, 1992, *Cooling our communities: a guidebook on tree planting and light-colored surfacing,* Washington DC: U.S. Environmental Protection Agency
6 Living Roofs and Walls: Technical Report Supporting London Plan Policy, Greater London Authority, February 2008
7 'Green Roofs: their existing status and potential for conserving biodiversiy in urban areas' was published by English Nature in 2003
8 http://livingroofs.org/

15 Transport

A refurbishment, even at the minor scale, comes about usually as a result of changes to the number of people utilising the facility, whether it be a floor being fitted out, or the reduction in space requirements arising from greater automation. Changes in the number of people or visitors arriving at the building can have dramatic impacts on the accessibility to the site and on the well-being of staff.

The provision of travel plans for refurbished buildings will help in understanding how transport improvements can be made, unlocking any particular issues, together with ensuring that any changes in staff or working patterns can be accommodated. This will also apply to deliveries to the site to ensure that access is not obstructed at key times during the day.

Mixed-use refurbishments (see Section 4.1: Better Utilisation of Buildings) can also help reduce the need to travel by providing work and living facilities together with access to retail and leisure facilities. Such an approach of reducing the need to travel, combined with the provision of alternative forms of transport, can reduce the number of car-parking spaces required and increase space for development. This is particularly key for edge-of-centre and suburban developments where a reduction of 75 per cent against the local average for commercial car parking can be achieved. In addition, significant reductions in travel-related carbon emissions can be achieved.

Sustainable Refurbishment, First Edition. Sunil Shah.
© 2012 Sunil Shah. Published 2012 by Blackwell Publishing Ltd.

Chapter Learning Guide

This chapter will review the transport issues facing organisations refurbishing properties and ways to effectively deal with them.

- Development and implementation of a travel plan;
- Use of a delivery-servicing plan to manage the supply-chain support services for businesses;
- Checklist of minimum requirements and good practice to implement sustainable material refurbishments.

Key messages include:

- Development of a travel plan for staff or tenants can help reduce travel to work by single-occupancy vehicles and therefore reduce the number of car-parking spaces required;
- Travel plans for deliveries can save money and reduce health-and-safety implications.

15.1 Developing a Travel Plan

The prime motivation for devising a travel plan is to promote a range of measures that will help to reduce journeys made by the single occupant of a private vehicle and encourage an increase in the use of walking, cycling and public transport.

The successful implementation of the plan will provide a range of benefits to the business, its staff, the local community and the environment. These benefits will include:

- Better health of staff and the encouragement of safe travel alternatives;
- Informed choice on travel alternatives to reduce the number of single-occupancy trips;
- Influence on the travel behaviour of employees;
- Better retention of staff;
- Reduction of vehicles in the car park;
- Improvement of the environmental image of the company;
- Reduction of traffic in the local area.

The travel plan needs to be focused on staff as there is a far greater success rate with this target group. Therefore, the majority of measures proposed are intended to encourage staff to vary, or change, their reliance on private-car travel. It is recognised that there is the potential to influence visitors' travel behaviour to a degree. The measures aimed at customers are directed towards increasing awareness of alternatives to private-car use through the display and promotion of information on noticeboards and leaflets.

Surveys should be undertaken within three months of occupation to identify travel patterns and provide a baseline. Annual surveys can be performed to ascertain the effectiveness of the travel plan through comparison with the initial staff survey. Whilst this approach provides evidence of the measures that are, and are not, working, it operates from a baseline that already includes a number of initiatives that have been implemented and will have influenced the initial staff survey. Where refurbishment takes place with an occupier in place, the initial staff survey should be conducted at least three months prior to the refurbishment's taking place.

The travel plan targets for non-single-occupancy car journeys will be dependent on the baseline survey data, but should look to set annual and longer-term targets for the two key indicators of single-occupancy vehicle and alternative travel means. The targets will therefore be agreed between the travel plan coordinator and the local authority following the initial staff-travel survey.

Travel Plan Initiatives

To ensure that the opportunities for modal shift can be realised, there are a number of measures that will need to be implemented to influence the modal choice for the journey to work and provide initiatives to enhance the dissemination of information to visitors regarding the opportunities for completing journeys by alternative travel modes to that of the car.

Provision of Travel Information Information relating to the potential for travel by non-car modes should be disseminated to staff, such as by the use of a noticeboard, located in a suitably accessible place, to increase staff awareness of the travel options available to them. A copy of the travel plan should also be made available to staff, via a letter circulated to staff, together with its promotion through an internal-communications route. The local authority will be made aware of the identity and contact details of the travel plan coordinator upon appointment.

Measures to Promote and Facilitate Cycling, Walking and Public-Transport Use Some initiatives to encourage the use of cycling and alternative forms of transport are provided below:

- Changing and washing facilities for employees;
- Secure lockers for employees;
- Secure and visible cycle-parking stands;
- Free transport home in the event of an emergency;
- Provision of information to staff on the bicycle, pedestrian and public-transport network routes, illustrated on maps;
- Provision of a shelter;
- Promoting a 'walking buddy' scheme for employees, similar to that for car sharing;
- Providing pedestrian access within, and into, the store, including paved areas, improved lighting, zebra crossings, and dedicated pedestrian areas;
- Establishing a bicycle-user group (BUG) for employees.

Leaflets detailing the health benefits of cycling and walking should be distributed and included as part of an employee 'starter' pack for those newly employed.

Car Sharing Car sharing is an effective means of reducing single-occupancy journeys to work. The practicalities of car sharing might be limited for some facilities due to varying working hours or shift working and, perhaps, due to the proportion of female employees who might not be willing to accept lifts.

The use of car-sharing-scheme websites for staff can be promoted through questionnaires and invitations to join the car-sharing scheme. Car-sharing leaflets should be made available on the staff-travel noticeboard, detailing the car-sharing proposals and how employees can get involved and participate in the scheme. A car-sharing registration form should be circulated to all those who indicate a willingness to participate.

The feasibility of the use of car-hire companies should be investigated, whereby pool cars or community vehicles could be made available to staff for business travel or even to travel home where necessary.

Travel Plan Coordinator Central to the success of the travel plan is the appointment of a travel plan coordinator. The appointed individual will be made known to the local authority upon inception and will be the main driving force behind the plan. The individual will work in conjunction with the local authority, the local community and other interested parties for the continuing progression of the travel plan.

The travel plan coordinator will need to be allocated sufficient time and resources to manage the plan effectively, with the following responsibilities:

- Oversee the development and implementation of the travel plan;
- Be the point of liaison with the local authority in respect of the plan;
- Collect data and information regarding local bus-service timetables, changes to services and amendments;
- Be the main point of contact for all staff requiring information regarding the travel plan;
- Ensure that all information regarding cycle routes/bus-service timetables are kept up to date and readily available;
- Be responsible for the maintenance of the travel plan noticeboard;
- Arrange for travel surveys to be undertaken where necessary;
- Manage the setting-up and running of the car-sharing database;
- Assist in monitoring the success of the travel plan.

15.2 Delivery Travel Plans

In a similar way to changes in staff levels, refurbishments can also bring significant changes to the level of business deliveries being made, particularly in multi-tenanted properties where centralised and tenant-specific deliveries are made.

A delivery travel plan (DTP) is a framework for identifying potential changes to business practices that enable an organisation to actively manage their deliveries, as a business version of a travel plan. The plan helps businesses to achieve efficiency gains or cost savings, improve operational safety, and reduce the environmental impact of transportation within the supply chain.

Typical delivery aspects to be included in such a plan include:

- Deliveries and collections;
- Servicing trips such as for maintenance of office machinery, boilers, and lifts;

- Cleaning and waste removal;
- Catering and vending.

A DTP is bespoke to the company undertaking it, and follows similar stages and activities to a travel plan in terms of management, implementation and monitoring. A DTP can sit alongside, and work in conjunction with, an organisation's travel plan to ensure that all transport operations associated with a company or site are efficient, cost-effective, and as sustainable as possible.

As with a travel plan, the nature of a DTP can vary from one or two simple activities to achieve quick gains, to a comprehensive programme of activities, set up to achieve the maximum benefits possible. Effective DTPs involve staff from a number of departments, working together, and with suppliers and contractors, to improve the efficiency of vehicle movements.

Managing Deliveries

Taking an active part in managing deliveries, and working with suppliers on delivery timings, frequencies, and other undertakings, can deliver a range of benefits:

- Less congestion on site, particularly in delivery or loading bays;
- Improved on-time supplier performance and more reliable deliveries;
- Improved productivity and staff efficiency;
- Reduced congestion in the locality;
- Reduced risks by scheduling delivery vehicles away from peak pedestrian or workforce movements;
- Being a better neighbour, by reducing the impact of delivery and collection vehicles in the neighbourhood;
- Better goods, product or material availability;
- Reduced 'Penalty Charge Notices' for operators by providing legal, loading-space availability at agreed times.

15.3   Transport Checklist

Table 15.1 provides a summary of the items to be discussed during a sustainable-refurbishment project, capturing where specific items are necessary, recommended or nice-to-have options.

Table 15.1: Transport checklist

Minimum Standard	Recommended	Nice to have
Develop and implement a green transport plan and have a displayed green transport policy	Design-in provisions for green transport, including primary transport options, bus stops/shelters, cycle facilities and pedestrian walkways	Aim to reduce single-occupancy travel to 60 per cent and limit users' reliance on cars
Parking capacity not to exceed planning agreement; additional parking available through off-site parking site	Priority car parking for green travel (e.g. electric, hybrid or multi-occupancy)	Covered parking canopies with PV/wind-generated lighting
Bicycle facilities for 5 per cent of workstations with shower and changing facilities	Provide bicycle security and changing/shower facilities for 10 per cent of building occupants close to occupied space	
Public-transport information provided on notice boards; safe and secure shelters provided for staff waiting	Shuttle buses provided for main routes to public-transport hubs during peak travel times	24–7 transport provided through shuttle buses and taxis
All fleet vehicles to be maintained in accordance with manufacturer's, industry and local requirements	All fleet vehicles to be high-efficiency with catalytic converters	All fleet vehicles to be non-petrol/diesel (e.g. electric, LPG)

Glossary

BIODIVERSITY The wide variability of organisms on earth and within an ecosystem. Maintaining biodiversity is necessary to preserve the health and survival of an ecosystem.

BIOMASS Living or recently-dead organic material that can be converted into use as an energy source, for example, wood, agricultural crops and waste, or ethanol and methane.

BREEAM (Building Research Establishment Environmental Assessment Method) An environmental assessment method for non-domestic buildings, widely used across the UK.

CARBON FOOTPRINT The total amount of greenhouse gases emitted through an activity or from a product, company or person either directly or indirectly. It is generally expressed in equivalent tonnes of either carbon or carbon dioxide.

CARBON-NEUTRAL The net zero-carbon emissions to the atmosphere from a product, company or person. Achieving carbon neutrality means measuring the carbon emissions, then balancing those emissions with carbon reductions or carbon offsets to reach net zero-carbon emissions.

CHP (Combined Heat and Power) The simultaneous generation of usable heat and power as a single process.

CLIMATE CHANGE A 'statistically significant' change in the 'average weather' that a given region experiences. Climate change on a global scale refers to changes in the climate of the Earth as a whole, including temperature increases (global warming) or decreases, and shifts in wind patterns.

CLOSED-LOOP RECYCLING   The process of utilising a recycled product in the manufacturing of a similar product or the re-manufacturing of the same product.

CRADLE-TO-CRADLE   A design philosophy that looks at the life cycle of a material or product.

CRC (Carbon Reduction Commitment) ENERGY-EFFICIENCY SCHEME   A mandatory cap-and-trade scheme in the UK that applies to large non-energy-intensive organisations in the public and private sectors.

DEC (Display Energy Certificate)   A DEC shows the actual energy usage of a building, the Operational Rating, and enables the public to see the energy efficiency of a building based on its energy consumption recorded from gas, electricity and other meters. DECs are required only for buildings that are occupied by a public authority, or for an institution providing a public service to a large number of people, thus being visited by a large number of people. DECs are valid for one year.

DEFORESTATION   Cited as one of the major contributors to global warming, it is the conversion of forested land to other non-forest land uses by the removal of trees and destruction of habitat.

ECOSYSTEM   A place having unique physical features, encompassing air, water, and land, and habitats supporting plant and animal life, including that of humans.

ENERGY ASSESSMENT   The preparation and issuing of an Energy Performance Certificate (EPC) and the accompanying Recommendation Report (RR), and the carrying out of any inspections, undertaken for the purposes of issuing the EPC or RR.

ENERGY AUDIT   An inspection, survey and analysis of energy flows in a building, process or system with the objective of understanding the energy dynamics of the system under study. Typically conducted to look at ways to reduce the amount of energy input into the system without negatively affecting the output(s).

ENERGY EFFICIENCY   The use of less energy to fulfil the same function or purpose; usually attributed to a technological fix, rather than to a change in behaviour; examples include better insulation to reduce heating/cooling demand, compact fluorescent bulbs to replace incandescent bulbs, or proper tyre inflation to improve vehicles' fuel consumption.

EPBD (Energy Performance of Buildings Directive)   The Directive on the Energy Performance of Buildings (EPBD) is the Directive 2010/31/EU (EPBD, 2003) of the European Parliament and Council on energy efficiency of buildings.

EPC (Energy Performance Certification)   All commercial properties require an EPC, which rates the energy efficiency of the building on a scale of A–G, where A signifies the highest level of efficiency. Unlike the DEC (which gives an Operational Rating), the EPC provides an asset rating of the building fabric and fixed services, based on a survey of the building.

FiT (Feed in Tariff)   A policy mechanism that is designed to encourage the adoption of renewable-energy technologies and to help accelerate the move towards security of supply and grid parity.

FOSSIL FUEL   Any fuel source, such as natural gas, fuel oil, or coal, that has a finite supply.

GLOBAL WARMING   An increased warming of the Earth's atmosphere that can be caused by an increase of man-made gases that trap the sun's heat. This can effect changes, such as sea-level rises, changes in rainfall patterns and frequency, habitat loss and droughts.

GREEN BUILDING   A comprehensive process of design and construction that employs techniques to minimise adverse environmental impacts and reduce the energy consumption of a building, whilst contributing to the health and productivity of its occupants. Common ways of measuring green buildings include BREEAM and LEED (Leadership in Energy and Environmental Design).

GREENHOUSE EFFECT   The trapping of heat within the Earth's atmosphere by greenhouse gases, such as carbon dioxide and methane, which accumulate in the Earth's atmosphere and act as a blanket, thus keeping heat in.

GREENHOUSE GAS (GHG)   These gases are so-named because they contribute to the greenhouse effect due to high concentrations of these gases remaining in the atmosphere. The GHGs that are of most concern include carbon dioxide (CO_2), methane (CH_4), and nitrous oxide (N_2O).

GROUND-SOURCE HEAT PUMP   A system of underground pipes that extracts naturally occurring heat from the ground and increases its temperature using a heat pump. The heat is then used to provide building heating or hot water.

HAZARDOUS WASTE   Waste that has the potential to cause harm to human health or the environment (e.g. contaminated soil).

LCA (Life-Cycle Analysis)   An environmental-impact tool used to compare the environmental performance of two or more scenarios. The LCA quantifies the potential environmental impacts, and can highlight areas in which to target improvements.

LCC (Life-Cycle Costing)   A procurement-evaluation technique that determines the total cost of acquisition, operation, maintenance and disposal of items, potentially being procured.

LEED™ (Leadership in Energy and Environmental Design)   A US-developed, green, building-rating system that encourages and accelerates the global adoption of sustainable green building and development practices.

PV (Photovoltaic)   Solar photovoltaic's (PV) collectors are arrays of cells containing a material that converts solar radiation into electricity.

RECYCLING   The series of activities, including collection, separation, and processing, by which materials are recovered from the waste stream for use as raw materials in the manufacture of new products.

RENEWABLE RESOURCES   A resource that can be replenished at a rate equal to, or greater than, its rate of depletion. Examples of renewable resources include corn, trees, and soy-based products.

SHW (Solar Hot Water)   Panels that collect the solar radiation from the sun and use it to heat a liquid inside the building. This liquid then transfers its heat to the hot water in a cylinder, for use in hot-water requirements, for example, domestic hot water.

SUSTAINABILITY   The most widely accepted definition comes from 'Our Common Future', Report of World Commission on Environment and Development, commonly called the Brundtland Report 1983. It says, 'Sustainable development is development that meets the needs of the present without compromising the ability of future generations to meet their own needs.'

TRAVEL PLAN (also known as a GREEN TRAVEL PLAN or COMMUTER PLAN)   A travel plan aims to promote sustainable travel choices (e.g. cycling, car sharing or better use of public transport) as an alternative to single-occupancy car journeys that may impact negatively on the environment, congestion and road safety. Travel plans can be required by planners when granting planning permission for new developments, often by way of planning conditions.

WASTE-TO-ENERGY   The burning of waste in a controlled-environment incinerator to generate steam, heat, or electricity.

Further Reading and Websites

The list below provides a full list of the various documentation and websites available for further information. It is by no means exclusive, but will certainly provide an initial starting point for further information to support the various topics discussed.

Life Cycle of Building/Products

Environmentally Preferable Products and Services (EPP) Scientific Certification Systems www.scs1.com
Forest Stewardship Council www.fsc-uk.info
Green Building Lifecycle Assessment www.eiolca.net/index.html
Green Seal www.greenseal.org

Commissioning

'A Retrocommissioning Guide for Building Owners' peci.org/Library/EPAguide.pdf
Building Commissioning Association www.bcxa.org
California Commissioning Collaborative, 'California Commissioning Guide: New Buildings' and 'California Commissioning Guide: Existing Buildings' www.cacx.org
Northwest Energy Efficiency Alliance www.betterbricks.com
PECI Commissioning Library peci.org/CxTechnical/resources.html

Energy Management and Climate Change

Building Research Energy Conservation Support Unit www.bre.co.uk/brecsu
Carbon Disclosure Project www.cdproject.net/

Carbon Trust www.thecarbontrust.co.uk

EnergyStar from US EPA www.energystar.gov

EU Emissions Trading Scheme europa.eu.int/comm/environment/climat/emission.htm

Kyoto Agreement, European Programme europa.eu.int/comm/environment/climat/kyoto.htm

Renewable Power Association www.r-p-a.org.uk

UN Framework Convention on Climate Change unfccc.int/

University of Southern California www.usc.edu/uscnews/stories/11608.html

Use of Resources

Envirowise www.envirowise.gov.uk

Institute of Civil Engineers Demolition Protocol aggregain.wrap.org.uk/demolition/the_ice_demolition_protocol/

UK Water www.water.org.uk/

Waste and Resources Action Programme (WRAP) has a range of activity on construction waste www.wrap.org.uk

Marketplace

Chartered Institute of Purchasing Supply www.cips.org

Environmentally Preferable Purchasing from US EPA www.epa.gov/opptintr/epp

Biodiversity

Business and Biodiversity www.businessandbiodiversity.org/

Sustainability Information

Business in the Community helps members improve their impact on communities and the environment www.bitc.org.uk

CSR Academy – a training and awareness site www.csracademy.org.uk

CSR Business information www.bsdglobal.com/issues/sr.asp

CSR News www.mallenbaker.net/csr/

EC Green Public Procurement ec.europa.eu/environment/gpp/index.htm

EIRIS Ethical Investment Research Service www.eiris.org/

Electronic Reporting Network for social, environmental, economic and corporate governance information www.one-report.com

EMAS www.emas.org.uk/

ENDS Report online www.endsreport.com

Environment Council, The www.the-environment-council.org.uk/

Environmental news and information www.edie.net/index.asp

EU CSR Programme www.eu.int/comm/enterprise/csr/index.htm

Forum for the future www.forumforthefuture.org.uk

FTSE4GOOD develops and maintains a series of global sustainable investment indices www.ftse4good.com

Global Reporting Initiative is compiling sustainable reporting guidelines www.globalreporting.org

Green Futures magazine www.greenfutures.org.uk
Greenpeace www.greenpeace.org
IISD The International Institute for Sustainable Development iisd.ca/
Intergovernmental Panel on Climate Change www.ipcc.ch
James Lovelock's Gaia philosophy www.ecolo.org/lovelock/whatis_Gaia.html
MEPI (Measuring Environmental Performance of Industry) project www.sussex.ac.uk/Units/spru/mepi/about/index.php
Next Step Consulting Ltd www.corporateregister.com provides the Corporate Register – Environmental Reports
Non-financial corporate reports www.enviroreporting.com/
Oxfam www.oxfam.org
Rainforest Action Network www.ran.org
Social Accountability International developers of SA8000 www.sa-intl.org/
Socially responsible investment www.socialfunds.com/
SustainAbility www.sustainability.com/
SustainableBusiness.com www.sustainablebusiness.com/
The Natural Step www.naturalstep.org.uk
The Worldwatch Institute Report www.worldwatch.org
UK Institute of Environmental Assessment and Management www.iema.net/
UN Food and Agriculture Organisation www.fao.org/
UN Global Compact www.unglobalcompact.org
UNDP Human Development Report 2001 hdr.undp.org/reports/global/2001/en/
WBCSD World Business Council for Sustainable Development www.wbcsd.ch/
World Economic Forum www.weforum.org/
World Resources Institute www.globalforestwatch.org
WWF www.wwf.org.uk/

Legislation

Australian Department of Environment and Heritage www.deh.gov.au/
European Environment Agency www.eea.eu.int/
European Union legislation and information europa.eu/documentation/legislation/index_en.htm
Hong Kong Environment Protection Department www.epd.gov.hk/epd
Montreal Protocol ozone.unep.org/new_site/en/index.php
NetRegs is a web resource to help small companies understand environmental legislation www.environment-agency.gov.uk/netregs
The Environment Agency is interested in waste and pollution on construction sites, SUDS and operational water use www.environment-agency.gov.uk
US Environmental Protection Agency www.epa.gov/epahome/

General Sustainable Buildings

ACCA's (Association of Certified Chartered Accountants) work on Corporate Social Responsibility and sustainability reporting www.acca.org.uk
Association of Environment Conscious Builders (AECB) www.aecb.net
Barbour Index www.barbourexpert.com
BetterBricks www.betterbricks.com

BRE – Building Research Establishment. See here for more information about BREEAM, EcoHomes, envest, Environmental Profiles, Green Guide, MaSC, post-occupancy evaluation, SMARTStart, SMARTWaste, whole-life costing and much more www.bre.co.uk

British Council for Offices (BCO) has produced advice on green roofs and fuel cells, and general guidance on 'Sustainability Starts in the Boardroom' and 'Sustainable Buildings are Better Business' www.bco.org.uk

British Property Federation (BPF) who have produced an Energy Guide for members www.bpf.org.uk

BSRIA – Building Services Research and Information Association www.bsria.co.uk

Building for Environmental and Economic Sustainability (BEES) National Institute of Standards & Technology www.nist.gov/el/

Building Owners and Managers Association www.boma.org

Business in the Environment www.business-in-environment.org.uk

Business Link www.businesslink.org/

CIBSE (Chartered Institute of Building Services Engineers) www.cibse.org

CIOB – Chartered Institute of Building www.ciob.org.uk

CIRIA – Construction Industry Research and Information Association. Also has details of Construction Industry Environmental Forum (CIEF) and Construction Productivity Network (CPN) www.ciria.org.uk

Commission for Architecture and the Built Environment (CABE) www.cabe.org.uk

Considerate Constructors Scheme is a code of practice for improved construction sites www.ccscheme.org.uk

Constructing Excellence: bringing together Construction Best Practice (CBP) and Rethinking Construction (Movement for Innovation, The Housing Forum, Local Government Task Force) www.constructingexcellence.org.uk/

Construction Industry Council (CIC) is the representative forum for the industry's professional bodies, research organisations and specialist trade associations. The Happold Lecture Series is available here, plus details about the Sustainable Development Committee www.cic.org.uk

Construction Industry Training Board ConstructionSkills www.citb.org.uk

Construction Resources www.constructionresources.com

Cool Roof Rating Council www.coolroofs.org

Co-operative Bank's Ethical Purchasing Index www.greenconsumerguide.com/epi.php

Design Quality Indicator www.dqi.org.uk

Environmental Building News www.buildinggreen.com

Environmental Design + Construction www.edcmag.com

Ethical Junction www.ethical-junction.org

EU Design Practices www.unep.or.jp/ietc/sbc/index.asp

FIT Buildings Network www.theFBnet.com

Global Alliance for Building Sustainability (GABS) Charter webapps01.un.org/dsd/partnerships/public/partnerships/51.html

Green Register of Construction Professionals www.greenregister.org

Greener Buildings providing information on LEED www.greenbiz.com/section/buildings-facilities

Greenguard Environmental Institute www.greenguard.org

House Builders Federation www.hbf.co.uk/

HVCA (Heating Ventilation Contractors Association) www.hvca.org.uk provides standard maintenance specification for mechanical services in buildings

Institution of Civil Engineers (ICE) www.ice.org.uk

International Institute for Sustainable Development www.iisd.org/

Leadership in Energy and Environmental Design www.usgbc.org/

McDonough Braungart Design Chemistry www.mbdc.com/

National House-Building Council (NHBC) www.nhbc.co.uk

Prefabrication technologies www.fabprefab.com

Prince's Foundation is involved in a number of projects, and espouse 'Enquiry by Design' www.princes-foundation.org

Rocky Mountain Institute: www.rmi.org

Royal Institute of British Architects (RIBA) who have a Sustainable Futures group www.architecture.com

Royal Institution of Chartered Surveyors (RICS) www.rics.org

Royal Town Planning Institute www.rtpi.org.uk

SchoolWorks is a body working towards better school refurbishment and management www.school-works.org

Sponge – network for young professionals in sustainable construction www.spongenet.org

Sustainable Buildings Industry Council www.sbicouncil.org/

Sustainable materials sourcebook www.greenbuilder.com/sourcebook/

Water, Waste and Environment suppliers www.water-waste-environment-marketplace.com/

Whole Building Design Guide www.wbdg.org/

WWF has launched an initiative for One Million Sustainable Homes www.wwf.org.uk/sustainablehomes

US Government

Building Deconstruction Consortium focuses on research, information dissemination and institutionalisation of deconstruction practices. To date, primary focus has been on military-base deconstruction www.buildingdeconstruction.org

Federal Green Building (FedGB) Listserv covers over 300 federal employees involved in green buildings www.epa.gov/greenbuilding

Federal Interagency Committee on Indoor Air Quality (CIAQ) www.epa.gov/iaq/ciaq/

Office of the Federal Environmental Executive: www.ofee.gov/

U.S. Department of Energy High Performance Buildings www1.eere.energy.gov/buildings/commercial_initiative/

U.S. Department of Housing and Urban Development's Guide to Deconstruction www.huduser.org/publications/destech/decon.html

U.S. Department of Interior's Guiding Principles of Sustainable Design www.nps.gov/dsc/d_publications/d_1_gpsd.htm

Index

Sustainable Refurbishment, First Edition. Sunil Shah.
© 2012 Sunil Shah. Published 2012 by Blackwell Publishing Ltd.

Keep up with critical fields

17841

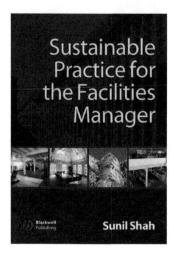
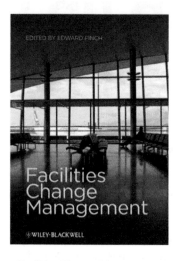
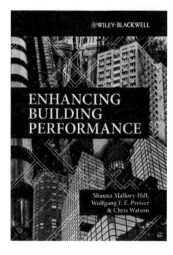

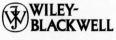

Printed and bound by CPI Group (UK) Ltd, Croydon, CR0 4YY

27/10/2024

14580193-0005